Fundame[illegible]

of

Matter and Fields

By

G S SANDHU

iUniverse books may be ordered through booksellers or by contacting:

iUniverse
1663 Liberty Drive
Bloomington, IN 47403
www.iuniverse.com
1-800-Authors (1-800-288-4677)

ISBN: 978-1-4401-3656-6 (sc)
ISBN: 978-1-4401-3657-3 (ebook)

Library of Congress Control Number: 2009926610

Printed in the United States of America

iUniverse rev. date: 4/17/2009

Preface

During the last century, all branches of applied science and engineering have shown tremendous progress. But in theoretical Physics, which had once been regarded as the Mother of all sciences, the situation is quite grim. The current thrust of front-line fundamental research in Physics, appears to have lost its momentum and is heading nowhere. Perhaps in the grand maze of the unknown, we have somewhere missed the right track and are now approaching a dead end.

The tragedy of 20th century physics, is the gradual shift in our focus from physical reality to abstract mathematical formulations, which are supposed to describe physical reality. The quest for some grand unified theory, no longer implies a quest for some fundamental theory that could fully explain separate phenomena of physical reality, and which could enable us fully understand and visualize all aspects of physical reality. But the meaning of Unification, in the present context, is the search for some grand mathematical structure, which could link the mathematical representation of all separate phenomena.

We have been steadily indoctrinated into believing that due to the complexity of physical reality, we can not demand mental visualization of basic phenomena in quantum mechanical world. For example, we are made to believe that interaction between two electrons takes place through mutual exchange of 'virtual' photons; but we are prohibited from demanding deeper understanding or mental visualization of such virtual particles. Further, we are also prohibited from attempting to mentally visualize the electron as a particle or its orbital motion around the proton. This sort of mass hypnosis in fundamental physics is seriously inhibiting significant major advancement in the field.

To fully comprehend and understand any physical phenomenon, we must demand mental visualization of such phenomenon. In this book I have presented a bold un-orthodox viewpoint that demonstrates the feasibility of representing whole physical phenomenon, involving matter and fields, in the form of orderly 'space-time' distortions or dynamic deformations and strains in the space continuum. Through study and analysis of dynamic deformations in the space continuum, all physical phenomenon as well as shape, size and structure of all matter particles and their associated fields, can be brought within our mental grasp and visualization. This viewpoint calls for a paradigm shift in fundamental physics for further advancements in this field.

G S Sandhu

March, 2009 gurcharn_sandhu@yahoo.com

To my wife and family

for their immense patience and continuing support

Contents

Introduction

This book has two parts. In the first part, various inadequacies of fundamental physics have been discussed, where a change in approach, a paradigm shift is called for. The second part studies the evolution of matter particles and fields from space-time distortions. As per the current theoretical approach, natural phenomenon is modeled on abstract mathematical notions, without recognizing the necessity for mental visualization and comprehension of the associated physical phenomenon. Mathematical models based on actual physical observations cannot obviate the necessity of a physical theory with causal linkages for logical explanation of the associated phenomenon.

In first two chapters it is brought out that whereas the metric scaling property is only associated with coordinate space, the physical measurable properties of permittivity ε_0, permeability μ_0 and intrinsic impedance Z_0 are only associated with physical space. Existence of aether medium has been under fierce debate for more than a century now. In reality, various notions of physical space, empty space, vacuum, aether and their modern reincarnation, the quantum vacuum, all mean the same entity – call it by any name. We will however, prefer to call this entity 'Elastic Space Continuum'. A significant point to be highlighted here is that just like the intrinsic impedance $Z_0 = \sqrt{\mu_0 / \varepsilon_0}$, the speed of strain wave propagation $c = \sqrt{1/(\mu_0 \varepsilon_0)}$ is also a measurable property of the physical space.

The curvature of space (and spacetime) is a sophisticated buzz word under the current paradigm of fundamental physics. However, the interpretation associated with the curvature of space is quite misleading. As per General Theory of Relativity (GR), gravitation induced 'change' in metric of space, induces a corresponding change in the value of Riemann tensor, which by '*convention*' implies a change in curvature of space. It is clarified in chapter 7 that any 'change' in the metric coefficients of space actually implies a change in arc length ds between two neighborhood points, which in turn implies a deformation of space. Any such deformation of space is mathematically linked with the strained state of the space continuum. Since gravitation induced deformation or strained state of the space continuum is technically reversible, it implies the association of elasticity property with the space continuum. Hence, we may regard the empty space as an **elastic space continuum**.

In chapters 3 to 6, various issues related to the Special Theory of Relativity (SR) have been discussed. It is shown that except for the mass energy equivalence, all other assertions and postulates of this theory are fundamentally wrong and misleading. Specifically it is shown that an absolute fixed reference frame, like the International Celestial Reference Frame (ICRF), can be constructed in any closed volume of space with finite matter content. The notion of infinitely many reference frames, flying around with relative uniform motion, with fictitious observers riding them, is basically redundant and misleading. The concept of absolute reference frame is extended to the Universal Reference Frame and an experimental technique has been outlined for practically establishing the same, just like the ICRF.

In chapters 8 and 9, it is shown that the notion of spacetime continuum or spacetime manifold is just a mathematical abstract notion and not a physical entity as assumed in the General theory of Relativity. In GR, localized mass energy content in a certain region of physical space (say the solar system) is '*supposed*' to influence the metric of whole spacetime, including the region of spacetime identified with the past time. But the transmission of 'influence' from the present to the past region of spacetime is logically impossible. Once we understand that spacetime is just an abstract mathematical notion, we cannot accept GR to be a physical theory. Even if we '*assume*' spacetime continuum to be a physical entity, it can be shown that the gravitation induced deformation of space leads to an incompatible set of strain components which cannot be valid due to physical constraints. At the most GR could be regarded as a mathematical model in which an abstract notion of spacetime manifold has been used as a graphical template with differential scaling of its space and time axes, to represent the particle trajectories as geodesic curves.

It is shown in chapter 10 that the famous uncertainty principle is strictly applicable for certain mathematical representations of the physical phenomenon and not to the phenomenon itself. However the uncertainty principle has scuttled our attempt to mentally visualize the physical phenomenon. For example, once we accept the shackle of uncertainty on our imagination, we cannot mentally visualize the orbital motion of electron around the proton. In chapter 11, it has been specifically demonstrated that without the uncertainty principle we can study the orbital motion of electron around proton in great detail. We can even simulate the photon emission phenomenon and visualize the orbital transitions. Further, it is shown that the so called total energy E is the total amount of energy already removed (emitted out) from the system and hence does not 'belong' to the system any longer. The negative potential energy V is the amount of energy released by the 'superposed' fields of the system of interacting particle and does not 'belong' to any one particle at any time.

While simulating the photon emission phenomenon during electron transitions from higher orbits to lower orbits, it emerges that the time required for emission of photon is just about 5 percent of the orbital time period in higher orbit. That means the emission of photon is not related to the physical oscillations of the electron and that its emission time is much shorter than L/c where L is the total length span of the photon along its flight path. In fact the photon gets emitted from the interacting fields of the electron and proton almost instantaneously. Based on this observation, a new hypothesis is proposed at chapter 12 regarding the variation of photon wavelength with the velocity of 'light source'. An alternative explanation for the null result of the famous Michelson–Morley experiment has been developed from this hypothesis.

Fundamental limitations of the Standard Model (SM) of particle physics have been discussed in chapter 13 of the book. In particular it has been shown that the exchange theory of interaction, which is the founding postulate of the SM, is fundamentally invalid due to the absence of any suitable physical mechanism for prior exchange of required information. Further, the SM has practically failed to provide adequate information on the characteristic shape, size and internal structure of any of the elementary particles. Hence, even if we retain the Standard Model as an excellent empirical mathematical model, we cannot afford to abandon the search for a more appropriate theory of elementary particles and their interactions.

Part II of the book mainly covers the hitherto unexplored field of space-time distortions or 'space dynamics' to analyze the shape, size, internal structure and mutual interactions of elementary particles. The study of space-time distortions or dynamic deformations in the space continuum, through a detailed study of corresponding time dependent displacement vector **U**, strain tensor **S** and stress tensor **T** fields in the elastic space continuum, may be termed as space dynamics. A detailed analytical study of dynamic deformations in the physical space continuum, through the time dependent displacement vector field **U** provides a more fundamental level of investigation into the workings of Nature, in comparison to the fields currently employed for the purpose. Apart from unification between electric and magnetic fields, the displacement vector field provides a unique mechanism to demarcate the physical boundaries of elementary particles.

In the Elastic Space Continuum, due to lack of discrete atomicity, we must assign the Poisson's ratio equal to zero. Also noting that there are no translational or rotational rigid body motions in the continuum, we find that the components of strain tensor $\mathbf{S}=[S^i_j]$ need not be symmetric.

The components of strain tensor referred to a coordinate system y^i, will therefore be related to the components of displacement vector $\mathbf{U}=[u^i]$ through the relation $S^i_j = \partial u^i/\partial y^j$ or $u^i_{,j}$. The Hooke's law of elasticity, relating the stress tensor $\mathbf{T}=[\tau^i_j]$ to the corresponding strain tensor components, takes the simple form $\tau^i_j = (1/\varepsilon_0)\ S^i_j$, where $1/\varepsilon_0$ is the elasticity constant for the continuum in appropriate units.

The equilibrium equations of elasticity in the Elastic Space Continuum turn out to be identical to the Maxwell's vector wave equation for the electromagnetic field as $\nabla^2\mathbf{U} = (1/c^2)\ \partial^2\mathbf{U}/\partial t^2$. The displacement vector **U**, strain tensor **S**, and the stress tensor **T**, are absolute entities and are invariant under coordinate transformations. These equilibrium equations, subject to appropriate boundary conditions, do not permit of any static (i.e. time invariant) strained state in the continuum and all permissible solutions in terms of displacement vector components u^i will be functions of space and time coordinates. The partial derivatives of u^i with respect to time t (more correctly ct) that is, $(1/c).\partial u^i/\partial t$ will constitute temporal strain components S^i_t in addition to the spatial strain components mentioned above.

The detailed study of any deformed or the stressed region of the Elastic Space Continuum primarily involves the detailed solution of the equilibrium equations subject to appropriate boundary conditions. In the deformed or stressed state of the Elastic Space Continuum, certain amount of strain energy will get stored in the region under stress. The strain energy density W or the energy of deformation per unit volume, at any point of the continuum, is a function of the intensity of strain at that point, and is an invariant. The strain energy density in the Elastic Space Continuum is given by $W = (1/2\varepsilon_0)$[Sum of squares of spatial and temporal strain components].

A closed region of Elastic Space Continuum in a strained state, satisfying the equilibrium equations & boundary conditions, may be termed as a strain bubble, provided the total strain energy content in this closed region is time invariant constant. Although the strain components at any point within the strain bubble are always functions of space and time coordinates, yet the strain energy density at that point may or may not vary with time. If the strain energy density at all points within a strain bubble is time invariant, the strain bubble is likely to be stable, otherwise unstable. The total strain energy content E_0 of a strain bubble will represent its 'rest mass' m_0 through the famous energy equivalence relation $E_0/c^2 = m_0$.

If the strain fields of two strain bubbles overlap in a certain region of the Continuum, then total strain components will be obtained by superposing the corresponding components of both the strain bubbles. Strain energy density and hence the total energy of the common field will be governed by the sum of squares of the resultant strain components. Interaction energy (E_{int}) or the conventional potential energy, of two such interacting strain bubbles is defined as the difference between the total strain energy with superposed strain fields (E_{sup}) and the sum of their separate strain field energies (E_1 and E_2) that is, $E_{int} = E_{sup} - (E_1 + E_2)$. A negative interaction energy or potential energy will imply release of a portion of the total strain energy of two interacting bubbles. The released energy may either transform into another strain bubble wholly or partly and emitted out of the system, or transform into kinetic energy of motion of the interacting strain bubbles.

At subatomic scale the primary constituents of matter, namely the electrons and nuclear particles are known to occupy an extremely small volume fraction of the order of 10^{-12} percent of the physical volume of any material body. These 'material particles' concentrated in such a small volume fraction of entire space, consist of so called 'elementary particles' and are essentially characterized by their 'mass', 'charge' and interaction properties. In the parlance of strain bubbles existing in the Elastic Space Continuum, the clusters of pure and composite strain bubbles depicting 'elementary particles' are essentially characterized by their 'strain energy content', 'strain wave fields' if any, and their interaction properties. In principle, there could be a large number of different types of strain bubbles occurring in the Space Continuum, that may be correlated with equally large number of stable and unstable elementary particles.

We have obtained particular solutions for the strain bubbles that correspond to well known stable particles namely electron, positron, proton, neutron and the photon. Strain bubble solutions that correspond to neutrinos and some of the unstable particles like mesons, have also been discussed. Specifically the electron, positron pair is found to be the lowest order spherically symmetric strain bubble solution of equilibrium equations. This strain bubble consists of standing strain wave core of about 1.61×10^{-15}m (1.61 fm) radius, surrounded by a radially propagating phase wave field extending up to infinity. The amplitude of these phase waves keeps diminishing with radial distance. These phase waves propagate outwards from the core at radial velocity c for the positron and propagate inwards to the core at radial velocity c for the electron. The effective strain components in this wave field are proportional to the 'root mean square' (rms) value of the amplitude of the corresponding strain waves. The unique characteristic features of these radial strain wave fields manifest in the unique charge property of these strain bubbles. Detailed computations of strain energy content show that almost 65 percent of the total mass energy of the electron is contained in its core and remaining 35 percent in its field.

The nucleon core is represented by one lowest order, cylindrically symmetric solution of equilibrium equations of elasticity in the Elastic Space Continuum. The strain energy density within the core region of this strain bubble is completely time invariant, implying overall stability of the nucleon. Detailed computations show that the nucleon core is of the shape of a right circular cylinder of 5.4 fm diameter and 3.1314 fm length. The familiar strong interaction between nucleons results from the physical overlap of their cores. The proton consists of a positron entrapped inside a nucleon core through strong interaction and moving around the nucleon core center within a radius of about 1fm. The neutron consists of an electron entrapped inside the outer periphery of a proton through strong interaction and moving around the core center within a radius of about 2.6 fm.

The strong interaction between two or more strain bubbles or particles is physically effected through the superposition of the strain fields in their cores. The so called range of strong interaction is small just because the physical spread of the core region is small. On the other hand the electrostatic as well as the electromagnetic interaction is physically effected through the superposition of the radial wave fields of the interacting particles. Since the electrostatic wave fields of all charge particles physically extend up to infinity, the range of electromagnetic interaction is said to be infinite. Similarly the gravitational interaction between two or more neutral particles is shown to be physically effected through the superposition of their neutral ripple wave fields, which too extends unto infinity.

Finally, the fundamental nature of the matter particles and their associated fields, as brought out in this book, represents only the proverbial 'tip of the iceberg'. Significant amount of further research work is called for in this direction. The Quantum Mechanics needs to be recast in terms of strain waves instead of probability waves. Finally, the detailed new understanding of the strong interactions among nucleons, positrons and electrons can be of great value in the planned development of nuclear fusion processes and controlled fusion devices for peaceful energy generation in the future.

PART I

Inadequacy of the Current Paradigm

1

Permittivity and Permeability Constants of Vacuum

1.1 Physical Units and Dimensional Analysis

All entities, parameters or quantities to which some physical dimension can be assigned, and which can be measured in specified units, can be broadly classified as physical entities, parameters or quantities. Measurement of a physical quantity involves comparison of the quantity to be measured with a reference standard. The reference standard of measurement is called a physical unit. Because of known relationships among various physical quantities involved in a natural phenomenon, a limited number of units, called base units, are sufficient for expressing all physical quantities. All other units which can be derived from these base units are called derived units. The physical quantities or parameters that are measured in base units are called the physical dimensions such as length [L], mass [M], and time [T].

All physical quantities can be expressed in terms of the physical dimensions [L], [M], [T], etc. All equations in physics express some sort of inter-relationships amongst various physical quantities associated with a particular natural phenomenon. In order to ensure compatibility of physical units on both sides of all equations, these equations must be dimensionally balanced. Dimensional analysis is a powerful conceptual tool applied in physics and engineering to check the plausibility of physical equations. It is also used to study the inter-relationships of various physical concepts and form reasonable hypotheses about complex physical situations that can be tested by experiment or by more developed theories of the phenomena. In short, dimensional analysis provides a linkage between physical concepts and their mathematical representation.

We can employ dimensional analysis to study the proportionality constants of permittivity ε_0 and permeability μ_0 associated with free space or vacuum. Specifically, we need to study whether these constants characterize any physical property of free space. Further, we need to examine whether these constants could be construed to imply the existence of an entity like aether or whether some physical properties could be associated with the notion of vacuum. We also need to examine whether there is some sort of arbitrariness in the choice of units and physical dimensions of various universal constants like ε_0 and μ_0 associated with vacuum.

Dimensionality of Proportionality Constants. For a detailed study of this problem, let us first consider a general system with inter-related parameters A_1, A_2, A_3, A_4 ... etc. Let a typical (physically observed) relationship of these parameters be written as,

$$(A_1^{a1}).(A_2^{a2}).(A_3^{a3}) = \text{constant} \tag{1.1}$$

where a1, a2, a3 etc. may be +ve or -ve digits or fractions. If the dimensions of parameters A_1, A_2, A_3 are well defined, then through dimensional analysis we can ascertain whether the constant in equation (1.1) is a dimensional number or a dimensionless number. Only if this constant is a dimensionless number, we can declare with certainty that this constant does not characterize the system, i.e. it does not represent any physical property of the system in addition to the ones already represented by A_1, A_2, and A_3. The magnitude of such a dimensionless constant may depend on the choice of units of the parameters A_1, A_2, and A_3. The choice of units here essentially implies the choice of scale and not a choice of dimensions of various parameters.

On the other hand, if the constant in equation (1.1) is a dimensional number, we can replace it with a dimensional parameter say B_1, so that equation (1.1) can be rewritten as,

$$(A_1^{a1}).(A_2^{a2}).(A_3^{a3}) = B_1 \tag{1.2}$$

In this case, the parameter B_1 will certainly characterize the system, that is, it will represent a physical property of the system; even though under given observational environment this parameter may remain constant in magnitude. The magnitude of such a dimensional parameter too is governed by the choice of scale in the units of the parameters A_1, A_2, and A_3. But the essence of parameter B_1 and its dimensions cannot be arbitrarily changed without simultaneously tampering with the essence and dimensions of parameters A_1, A_2 etc. On the whole, a unit system is highly inter-related and dimensions of any one parameter cannot be arbitrarily changed without affecting all other dependent parameters. It is, however, possible that parameter B_1 may be a lumped parameter, that is, it may consist of two or more physical parameters which remain constant under given observational environment.

Examples of Dimensional Proportionality Constants. As a typical example, let us consider the case of Boyle's Law for an ideal gas at constant temperature. The relation between pressure p and volume V is written as,

$$p.V = \text{Constant} \tag{1.3}$$

It can be easily seen through dimensional analysis that this constant in equation (1.3) is a dimensional constant. Therefore, we can replace it with a dimensional parameter or with a group of dimensional parameters R.T that characterize the gas in its current state. This leads us to the perfect gas law or the equation of state.

$$p.V/T = R \tag{1.4}$$

where T is the absolute temperature and R is the characteristic gas constant. Here again, R is a dimensional constant, which can be further split into m (mass of the given quantity of gas in moles) and R_0 (the universal gas constant). Therefore, equation (1.4) reduces to:

$$p.V/(m.T) = R_0 \tag{1.5}$$

Here too, R_0 is a dimensional parameter that characterizes fundamental features of an ideal gas. Magnitude of R_0 varies in different unit systems but it does not imply that units of R_0 can be chosen arbitrarily. The essence of R_0 does not change with the change in unit system. However, we cannot extract the true essence of R_0 from the perfect gas law or all the associated experimental data. To fully understand as to how R_0 characterizes an ideal gas system, we need a kinetic theory of gases, which tells us that R_0 is a composite parameter consisting of N_0 (Avogadro's number) and k_B (Boltzmann's constant).

1.2 Dimensionality of ε_0 and μ_0 Constants

Let us take up the case of permittivity and permeability. We have following three relations involving the parameters ε_0, μ_0 and the charge Q which enable us to fix their units.

$$F = [1/(4.\pi.\varepsilon_0)].(Q^2/r^2)$$

or $$\varepsilon_0 = [1/(4.\pi.F)].(Q^2/r^2) \tag{1.6}$$

where F is the force acting between two equal charges Q separated by distance r.

And $$F/L = (\mu_0/2\pi).(I^2/d)$$

or $$\mu_0 = 2\pi.(F/L).(d/I^2) \tag{1.7}$$

where F/L is the force acting per unit length of two parallel conductors separated by distance d and carrying current I. Since the units of time are well established, we need to establish the units of only one of the two parameters Q and I, the other will get fixed automatically.

Also, $$\sqrt{\mu_0 \varepsilon_0} = 1/c \tag{1.8}$$

where c is the velocity of light with well-established units. On the basis of these three equations (1.6), (1.7) and (1.8), the units of three parameters ε_0, μ_0 and Q (or I) can be fixed satisfactorily. The units thus established, will be self-consistent and mutually compatible. We cannot arbitrarily change the units of any one of these parameters without affecting the units of other two. In MKSA system, units of parameters ε_0, μ_0 and I (hence Q) have been fixed this way and they satisfy all of the above mentioned three equations. We may once again highlight the fact that the proportionality constants in equations (1.6), (1.7) and (1.8), being dimensional parameters, play an extremely important role in characterizing the system, in characterizing the entity called free space, or vacuum or the aether.

It will not be out of place to mention here that in Gaussian CGS system of units, a bold attempt had been made to fix the proportionality constant in equation (1.6) to unity, apparently to give a simpler look to various equations. It is very interesting to find out how exactly it was done. Essentially the two parameters Q and ε_0 in equation (1.6) are lumped into one new parameter, (say) Q_g with units of statcoulomb. Correspondingly the unit of current in this system is also named statamp. Replacing $(1/(4.\pi.\varepsilon_0)).Q^2$ with Q_g^2 in equation (1.6) and $(1/(4.\pi.\varepsilon_0)).I^2$ with I_g^2 in equation (1.7) we get,

$$F = (Q_g^2/r^2) \quad \text{Or} \quad (1/F).(Q_g^2/r^2) = 1 \tag{1.9}$$

and

$$F/L = (\mu_0/2\pi).(I_g^2/d).(4.\pi.\varepsilon_0) = 2.(\mu_0.\varepsilon_0).(I_g^2/d)$$
$$= (2/c^2).(I_g^2/d)$$

or

$$(F/2L).\ (d/I_g^2) = 1/c^2 \tag{1.10}$$

Units of various other parameters in Gaussian CGS system are then fixed such that they are consistent with equations (1.9) and (1.10) and mutually compatible. Here it is very important to note that the dimensions of charge Q in MKSA system are entirely different from the dimensions of charge Q_g $(=Q/\sqrt{(4.\pi.\varepsilon_0)})$ in Gaussian system and hence they do not represent the same physical quantity. However, most often this distinction is overlooked.

1.3 Significance of Dimensional Constants

Finally, let us consider one more example relevant to this discussion. Take a long thin metallic rod and subject it to load test. We find stress is proportional to strain.

$$\text{stress/strain} = \text{constant} \tag{1.11}$$

From dimensional analysis we find that this proportionality constant is a dimensional parameter Y known as the Young's Modulus. Obviously here Y is not just another arbitrary constant depending on arbitrary choice of units. Since the dimensions of stress and strain are well established, only magnitude of Y can vary with the variation of scale factor of different units. The dimensions and hence the essence of Y remains invariant. Y represents a very important property (the elasticity) of the material of the test piece under consideration.

Now let us conduct another test on the given rod. If we measure the mass and volume of any piece of this rod, we find that mass is proportional to the volume.

$$\text{mass/volume} = \text{constant} \tag{1.12}$$

Here again the proportionality constant is a dimensional number. We, therefore, replace it with a dimensional parameter ρ. The density ρ also represents a very important property (the inertial property) of the material of the test piece under consideration and is not an arbitrary constant. Finally if we measure the velocity c_l of longitudinal strain wave propagation in the given test piece, we find it related to Y and ρ as,

$$\sqrt{Y}/\sqrt{\rho} = c_l \tag{1.13}$$

This illustrates the importance of the dimensional proportionality constants, which cannot be discarded by just branding them as arbitrary proportionality constants.

1.4 Significance of Permittivity and Permeability Constants

Now coming back to the case of permittivity ε_0 and permeability μ_0 of the so-called empty space or vacuum or the old aether medium, let us rewrite equation (1.8) as,

$$\frac{\sqrt{1/\varepsilon_0}}{\sqrt{\mu_0}} = c \tag{1.14}$$

Comparison of equations (1.13) and (1.14) shows a remarkable similarity of the two cases. Velocity c_l represents the velocity of strain wave propagation in the material medium and velocity c represents the velocity of electromagnetic wave propagation in the so-called empty space or vacuum or the aether medium. Therefore, the electromagnetic wave propagation may be compared with a strain wave propagation. In this comparison $(1/\varepsilon_0)$ may be seen to be identical to the elasticity property Y and (μ_0) may be seen to be identical to the inertial property ρ.

Therefore, it stands to reason that we must strive very hard to unravel and to comprehend the deeper significance of the proportionality constants ε_0 and μ_0 associated with the entity called empty space or vacuum or the aether medium or identified by any other name. It is said, "Rose by any other name will smell as sweet". However, we would like to call this entity the 'Elastic Space Continuum'. Hence, we need to examine and comprehend the physical properties associated with empty space or vacuum or the Elastic Space Continuum.

2

Empty Space, Aether and Vacuum

2.1 The Coordinate Space

The crisis of modern physics can be attributed to a wide spread mix-up between the abstract mathematical notions and the physical concepts of space. Therefore, we need to bring out a clear distinction between the two notions of space in vogue. The first notion is that of a mathematical or coordinate space and the second is that of physical space. Let us consider them one by one.

The cardinal idea responsible for the invention of coordinate systems by Descartes consists of the assumption that to each real number there corresponds a unique point on a straight line. The association of the set of points P on coordinate line X with the set of real numbers x, constitutes a coordinate system of the one-dimensional space, once the notion of certain unit length has been defined. The one-to-one correspondence of ordered pairs of numbers with the set of points in the plane X^1X^2 is the coordinate system of the two-dimensional space. The extension of this representation to points in a 3-dimensional space is obvious. With predefined notion of unit length, the essential feature of it is the concept of one-to-one correspondence of points in space with the ordered sets of real numbers. The predefined notion of unit length or scale for different coordinate axes constitutes the metric of space for quantifying the position measurements of the sets of points in this coordinate space. [1]

We define a space (or manifold) of N dimensions as any set of objects that can be placed in a one-to-one correspondence with the ordered sets of N numbers $x_1, x_2, \ldots.., x_N$ such that $0 \leq x_k <$ infinity. Any particular one-to-one association of the points with the ordered sets of numbers $(x_1, x_2, \ldots.., x_N)$ is called a coordinate system and the numbers $x_1, x_2, \ldots.., x_N$ are termed the coordinates of points in the coordinate system. In all coordinate spaces that are metricized, we associate the notion of unit length along all coordinate axes and a metric tensor g_{ij} with each coordinate system. All essential metric properties of a metricized space are completely determined by this tensor.

A necessary condition for the equality of mixed partial derivatives of a function u(x,y) is that u(x,y) be of class C^2; that is, the function together with its first two partial derivatives are continuous. But this restriction alone is not sufficient to insure the equality of mixed covariant derivatives. It can be shown that, if the order of covariant differentiation is to be immaterial, our tensors must be defined over a particular metric manifold X for which a certain tensor of rank four, made up entirely of the g_{ij} components, vanishes. This tensor known as the Riemann - Christoffel tensor R_{ijkl} plays a basic role in many investigations of differential geometry, dynamics of rigid and deformable bodies, electrodynamics, and relativity.

2.2 The Physical Space

The notion of physical space implies the spatial extension of the universe wherein all material particles and all fields are embedded or contained. The true void between material points is in essence the physical space. Any region of space which is devoid of any material particle is known as empty space or free space. It is important to note here that the coordinate space, along with its scale or metric, is our 'human' creation intended to facilitate the quantification of relative positions of material particles and fields. The existence of physical space does not depend in any way on the existence or non-existence of coordinate systems and coordinate spaces. Of course, for the study and analysis of physical space and the material particles and fields embedded in it, we do need the structure of coordinate systems and coordinate spaces as a quantification tool. The most significant point to be highlighted here is that whereas the metric scaling property is only associated with coordinate spaces, the physical properties of permittivity, permeability and intrinsic impedance are only associated with the physical space. Even though the notion of material particles and 'fields' being embedded or contained in the physical space, is generally accepted yet, the detailed mechanism involved in this embedding is not known. Obviously, such a mechanism must involve the known physical properties of free space.

2.3 Notion of Aether, Vacuum or Quantum Vacuum

In 19th century Physics, the notion of an all-pervading medium called aether was considered a necessity for distinguishing the concept of physical space from that of the coordinate space. However, some self-contradicting properties had to be ascribed to this aether. It was supposed to be an extremely thin medium to enable resistance free motion of solid bodies through it. At the same time it was required to be an elastic solid to enable the transverse (light) wave propagation through it. This was essentially due to the fact that matter and aether medium were regarded as two separate, independent entities. Maxwell's development of the electromagnetic theory of light, the null result of Michelson-Morley experiment and Einstein's special theory of relativity, apparently rendered the notion of aether superfluous. Electromagnetic field was granted an independent status, capable of independent existence just as matter. The relativity theories just brushed aside the very necessity of aether by declaring through its postulates that the coordinate space is de-facto the physical space.

Now, there is a growing realization in scientific circles that matter and electromagnetic field, both appear to have a common origin in empty space or vacuum. There is also a notion of vacuum energy and the phenomenon of creation, annihilation and transmutation of unstable elementary particles occurring in vacuum. As per the current viewpoint, empty space or vacuum no longer represents 'nothingness' but is supposed to be the seat of, or supporter of, all ultra-microscopic phenomenon of nature. This entity representing the old 'empty space' or vacuum has now been assigned a modern name of 'quantum vacuum'. This reincarnation of poor old aether certainly looks much more sophisticated and acceptable. The 'quantum vacuum' is thought of as - "a seething froth of real particle-virtual particle pairs going in and out of existence continuously and very rapidly". The quantum vacuum is considered to be a dynamic condition of equilibrium in which this reversible process is occurring everywhere extremely quickly. In reality however, all these notions of physical space, empty space, vacuum, aether and their modern reincarnation the quantum vacuum, all mean the same entity – call it by any name. Hence, the physical space continuum, referred by any other name like vacuum or quantum vacuum or aether will still have the same physical properties. We will however, prefer to call this entity 'Elastic Space Continuum'.

2.4 Physical Properties of Vacuum or Aether

Fundamental properties of this vacuum or empty space are represented by the following dimensional parameters.

Permittivity of free space = $\varepsilon_0 = 8.854 \times 10^{-12}$ Coulomb^2/N. m^2

Permeability of free space = $\mu_0 = 1.257 \times 10^{-6}$ Weber / Amp. m

$= 1.257 \times 10^{-6}$ N / Amp^2

Speed of propagation of EM waves in vacuum = $c = 2.998 \times 10^8$ m/s

The intrinsic impedance of vacuum = $Z_0 = 377$ Ohms

These four parameters are dimensional constants and hence represent fundamental physical properties of vacuum. The speed c of propagation of electromagnetic disturbances is governed by the permittivity ε_0 and permeability μ_0 constants associated with the empty space or vacuum. Since these four parameters are inter-related, only two of these are independent.

$$c=\sqrt{1/(\mu_0\varepsilon_0)} \qquad \text{and} \qquad Z_0=\sqrt{\mu_0/\varepsilon_0}$$

It needs to be strongly emphasized here that the parameter c given above represents a fundamental physical property of vacuum or aether and not a property of photons or EM waves. Just as the speed of stress/strain waves in a material media depends entirely on the physical properties of that media, the speed of photons in a transparent material also depends on the relative permittivity and permeability of that media. Similarly Z_0 represents a fundamental physical property of vacuum or aether and not a property of the electro-magnetic (EM) field. Further it is interesting to note that μ_0 can be replaced with Z_0/c and $1/\varepsilon_0$ can be replaced with $c.Z_0$ in all relations involving μ_0 or ε_0. These parameters are quite routinely measured experimentally and are universally well known. Since the intrinsic impedance of vacuum, Z_0 is 377 Ohms, it does give the impression that perhaps the aether or vacuum is primarily the seat of electrical phenomenon of nature. However the propagation of transverse waves in a continuous media is essentially a feature of mechanical phenomenon. Hence, we need to reinterpret these physical properties in mechanical terms.

Mechanical Interpretation of Physical Properties of Vacuum. Now we have two different notions of vacuum or physical space; one with dimensional properties of ε_0, μ_0, c and Z_0 and the second with fundamental dimensional properties of elasticity and inertia to enable transverse wave propagation through it. In order to establish a correlation between these two notions of vacuum or the space continuum, we need to postulate the equivalence of their fundamental properties. Thus we assume that the parameter $1/\varepsilon_0$ (or $c.Z_0$) represents the elastic constant and μ_0 (or Z_0/c) represents the inertial constant of the physical space continuum or vacuum. Appropriate physical dimensions can be assigned to these parameters through dimensional analysis. The plausibility of this assumption is confirmed by the fact that square root of (elastic constant / inertial constant) represents the velocity of strain wave propagation in an elastic continuum and the square root of $((1/\varepsilon_0)/\mu_0)$ also represents the velocity of transverse electromagnetic wave propagation in vacuum. Further, since the light waves propagate as transverse waves, the fluid characteristics of aether or vacuum are totally ruled out. Ideally, this fact should get highlighted in a more appropriate name of physical space or vacuum. That is, we might refer to the vacuum or empty space in such a way so as to highlight its elastic properties.

With this correlation between different notions of physical space or vacuum, it is formally established that the aether, vacuum, physical space, empty space and the quantum vacuum, all represent one and the same entity which is characterized by its elastic and inertial properties. To make this entity a little more representative of its elastic properties, which are so very necessary for supporting transverse electromagnetic wave propagation, we may assign a more appropriate name to this entity - 'The Elastic Space Continuum'.

2.5 Detection of Aether, Vacuum or Elastic Space Continuum

One crucial question that needs to be answered is:- how exactly do we detect the existence of aether or vacuum or the elastic space continuum? The detection and measurement of any physical entity actually involves the detection and measurement of some of its characteristic attributes. Such characteristic attributes of the Aether or the Elastic Space Continuum are ε_0, μ_0, c and Z_0, which are well known and have been measured quite precisely.

However, there is one special feature of this detection on which scientists had persistently focused their attention for a long long time. Since the aether or vacuum represent the physical space within which all material particles and fields of the universe are embedded, we should be able to define a special coordinate reference frame which is at rest in this physical space or aether. This special reference frame is termed as the absolute or universal reference frame. In principle we should be able to refer the motion of all material particles in the universe to this absolute reference frame and all such motion will then be termed as absolute motion. It has been argued that to detect the existence of aether, we should be able to detect this absolute reference frame.

The famous Michelson-Morley (MM) experiment[2] was one of many such attempts made to detect this absolute or universal reference frame, which however, did not succeed. Since the speed of propagation of light c is a physical property of aether or vacuum, it must be a universal constant in this absolute or universal reference frame. Obviously therefore, if an observer is in motion in the absolute reference frame, the speed of propagation of light must appear to be different from c to this observer. It was this change in the speed of propagation of light which was attempted to be measured in the MM experiment (MMX) through the expected variation in the interference fringes. The actual MM experiment yielded null result which was wrongly interpreted as a proof of non-existence of the absolute or universal reference frame. However, it can be logically explained that the null result of MM experiment was due to the invalidity of an implicit assumption that frequency of emission of light photons is not influenced by the motion of source in the absolute reference frame.

Independent Status of Matter and Fields. Let us now examine the next pertinent question as to how exactly particles of matter could move through an elastic space continuum without any resistance. For this we need to view material particles as a sort of lumped up strain energy, or, a sort of localized strain wave packets. For the elastic space continuum, the equilibrium equations of elasticity can be shown to be identical to the vector wave equation. Particular solutions of these equilibrium equations as functions of space-time coordinates, satisfying appropriate boundary and stability conditions within a bounded region, can be shown to represent various strain wave fields and strain wave packets. The electromagnetic field as well as all other forms of energy and matter can be shown to exist in the the elastic space continuum as strain wave fields or strain wave packets. The energy density associated with these stress/strain waves in any particular region of the space continuum will be proportional to the square of the intensity of such waves. The matter particles essentially exist in this elastic space continuum as packets of standing strain wave oscillations whose total strain energy remains conserved in the absence of any interaction with other strain waves or packets. Hence matter and EM field do not have any independent existence separate from the physical space.

To view the motion of material particles in the elastic space continuum, we may consider a rough analogy of a boat moving on water or as a tornado moving in the atmosphere. Just as a moving boat is always accompanied by surface waves in its vicinity, a matter particle moving in the space continuum will always be accompanied by a localized strain wave field, something like De Broglie waves. Kinetic energy of a particle in motion in the elastic space continuum can be viewed as the strain energy stored in the accompanying strain wave field.

3

Inertial Property and Time Measurements

3.1 Notion of Inertia

Inertia is an inherent property of all forms of energy and mass is the quantitative measure of inertia. The annihilation and pair production of some of the elementary particles has shown that the elementary particles could be viewed as specially lumped up or locally entrapped forms of energy. Therefore, it is quite reasonable to deduce that Inertia must be an inherent property of all forms of energy, especially the entrapped energy. The quantitative measure of inertia of a small energy content dE may be given by its equivalent mass content dm through the well known relation

$$dm = dE / c^2 \qquad (3.1)$$

where c is the speed of light in vacuum.

To elaborate this point, let us consider a small packet of energy (with energy content E_0) at rest in the Barycentric Celestial Reference Frame (BCRF). This packet of energy E_0 will be generally perceived as a particle of mass $m=E_0/c^2$ at rest in BCRF and hence possessing zero kinetic energy in this frame. Let us now consider the motion of this energy packet or the particle such that it moves at velocity v in BCRF and is said to possess kinetic energy $E_k=m.v^2/2$. As per de Broglie's hypothesis, all microscopic particles display a wave-like nature while in motion. The de Broglie relations show that the wavelength λ of the waves associated with a particle in motion, is inversely proportional to the momentum p of the particle ($\lambda = h/p$) and the frequency f is directly proportional to its kinetic energy ($f = E_k/h$).

What we understand from this is that kinetic energy E_k of a moving particle is contained in the pilot wave associated with the moving particle. Putting it the other way round, to make the particle move, we have to supply certain amount of energy E_k which gets stored in the pilot wave associated with the moving particle. The micro process of supplying energy E_k to the particle involves the notions of force, resistance to motion and inertia.

It implies that the inertial property of a particle is associated with the development of pilot wave, containing energy content E_k, whenever the state of the particle is changed from rest to motion. If there were no pilot wave associated with a micro particle in motion, there would have been no need to supply additional energy E_k to move the particle and hence there would have been no property of inertia associated with any particle or energy packet. The phenomenon of a pilot wave getting associated with any particle in motion, may be attributed to some special characteristics (like permittivity ε_0 and vacuum impedance Z_0) of free space. But the property of inertia can only be attributed to the development of pilot waves (containing certain energy content E_k) associated with the motion of any particle or energy packet.

3.2 Dynamic Mass

From this inertial property of all forms of entrapped energy, we can derive the notion of dynamic mass and develop its quantitative relationship with the rest mass. Let a material particle P be at rest in some center of mass (CoM) fixed reference frame like BCRF and let its rest mass in this frame be m_0. When at rest, the kinetic energy of this particle P will obviously be zero. Now let us assume that the particle P is set in motion through application of a constant force **F**. Further, at an instant of time t, let the instantaneous velocity of P be **v** with corresponding kinetic energy content E. Since the energy content E will also exhibit the inertial property, let the quantitative measure of total inertia of P at the instant t be given by m, which may also be referred as the dynamic mass of the particle. If during a small interval of time dt the particle traverses a small distance **ds** and gains a small amount of kinetic energy dE then the following relations will hold.

$$\mathbf{v} = \mathbf{ds}/dt \tag{3.2}$$

$$dE = \mathbf{F}.\mathbf{ds} \tag{3.3}$$

From Newton's second law of motion

$$\mathbf{F} = d(m\mathbf{v}) / dt$$

$$= m.\, \mathbf{dv}/dt + \mathbf{v}.\, dm/dt \tag{3.4}$$

From equations (3.3) and (3.4)

$$dE = m.\, (\mathbf{dv}/dt)\, .\, \mathbf{ds} + \mathbf{v}.(dm/dt)\, .\, \mathbf{ds}$$

$$= mv\, .\, dv + v^2.\, dm \tag{3.5}$$

And from equations (3.1) and (3.5) we get,

$$dm = (mv/c^2)\, .\, dv + (v^2/c^2).\, dm \tag{3.6}$$

Let us make a substitution x= v/c in equation (3.6) so that dx = dv/c and

$$dm = mx\, .\, dx + x^2\, .\, dm \tag{3.7}$$

Or, $(1-x^2)\, dm = mx\, .\, dx$

Or, $dm/m = (x/(1-x^2)) . dx$ (3.8)

This on integration yields,

$$m/m_0 = (1-x^2)^{-1/2}$$

or $$m = m_0/\sqrt{1-v^2/c^2}$$ (3.9)

This is a standard relation for the dynamic mass of a particle in motion. Here, it is important to note that the derivation of dynamic mass m in terms of rest mass m_0 did not involve special relativity. Instead, this derivation is entirely based on the inertial property of all forms of energy, including kinetic energy. Similarly, all dynamic relations of special theory of relativity (SR) can be shown to be resulting from the inertial property of all forms of energy.

3.3 Notion of Time

What is time? This issue has been debated and pondered over since ages. So far the best concept of time has been the intuitive concept; what we all feel by way of our common sense and experience. But still, there is no unique definition of time. The most common definition may be assumed as, 'Time is what we measure with clocks – all sorts of clocks'. But this definition does not throw any light on the fundamental basis or the origin of the notion of time. For the physical description of natural phenomenon, time (T) is regarded as one of the physical dimensions, just like mass (M) and length (L). However, due to the over bearing dominance of mathematics during the 20th century, some mathematicians have started propagating the idea of equivalence of the dimensions of length (L) and time (T). This has shattered the intuitive notion of time and led to a lot of confusion and controversy over the issue. Therefore a critical review of the fundamental basis or the origin of the notion of time has become all the more important now than ever before.

3.4 Cyclic Natural Phenomena

It is a well known fact that all natural phenomena, in the observable Universe, is intrinsically dynamic with a dominant feature of continuous *change***.** To explore the fundamental basis of the notion of time, let us conduct a thought experiment. Let us take a mental snapshot of the natural phenomenon and store it in memory. Let us take another snapshot of the same natural phenomenon and again store it in memory. Now we compare the two snapshots to see whether they are identical or there are noticeable *changes* in the second snapshot as compared to the first. Let us imagine we take a series of such mental snapshots and keep storing them in the memory in a systematic order. After examining a large number of such consecutive snapshots, we categorize various types of changes into a few distinguishable sets of changes. Further, let us assume that we concentrate our attention to only those changes which could be quantified through a measurable parameter.

Let us imagine that type A change is quantified through a measurable parameter 'a', type B change is quantified through a measurable parameter 'b' and so on. By studying a series of large number of successive snapshots we can analyze the nature of changes in parameters a, b, c, etc. While some of these parameters may vary randomly, some others may increase or decrease monotonically over a series of snapshots. Further, let us concentrate our attention to only those changes where the measurable parameters a, d, h, etc. continuously vary between specific limits say (a_1, a_2), (d_1, d_2), (h_1, h_2), etc. over a series of snapshots. We designate such changes as cyclic changes.

3.5 Cyclic Changes and Measure of Time

Suppose in a particular series of successive snapshots, parameter 'a' goes through N_a cycles of changes, parameter 'd' goes through N_d cycles of changes and so on. We can then compare the two sets of changes by saying that while parameter 'a' completes one cycle of change, parameter 'd' completes (N_d/ N_a) cycles of change. And this comparison between different sets of changes gives rise to the notion of comparatively slow or fast changes and hence to the notion of **time as a measure of rate of change.**

After a thorough analysis of all such cyclic changes, we may agree to designate the cyclic changes of parameter 'a' as a reference scale for comparing all other changes. In particular, we may call one cycle of change in parameter 'a' as one reference unit or just one unit of 'time'. Now when we say that parameter 'd' takes 10 units of time to complete its one cycle of change, it implies that in a series of successive snapshots, while parameter 'd' completes one cycle of change, parameter 'a' completes 10 cycles. Thus the notion of time helps us to compare different sets of change against one reference set.

Measure of Time. In Nature, there are a large number of physical processes, which undergo cyclic changes. Depending on the consistency of such cyclic changes and the convenience of their measurement, we may select any one of them as our reference scale for relative measurement of change or the reference scale for time. The angular position of a planet in orbit around the Sun, the angular position of an electron orbiting around the nucleus of an atom, the position of a pendulum oscillating about a mean, the vibrations of many mechanical, electro–mechanical and electro-magnetic systems are all examples of physical processes that undergo cyclic changes. Any such system or process could be adopted as a reference scale for relative measurement of change or measurement of Time. In general, the study of natural phenomenon invariably involves the comparative study of various changes. For this comparative study, we need to use a reference scale, or more correctly a reference time scale, for relative measurement of change or for measurement of time. Hence the Time, as a relative measure of change, is an extremely important parameter in the study of an essentially dynamic physical Universe.

In real life, the most common reference time scales adopted are the orbiting cycle of earth around the sun or the rotation cycle of earth around its axis. The cyclic oscillation period of a standard pendulum has also served as a reference time scale over the ages. Currently the cyclic oscillation period of certain opto-electrical processes serve as precision reference time scales. At present the official SI definition of a second is - "The duration of 9,192,631,770 periods of the radiation corresponding to transition between the two hyperfine levels of the ground state of cesium 133 atom at a temperature of 0 Kelvin."

However, the most important aspect of the measure of time is that any specific time interval 'dt' consists of a ratio of the elapsed number of cyclic oscillation periods to the number of oscillation periods defined in the reference time scale adopted. This measurement process for time is in essence similar to the one adopted for length measurements - as the ratio of certain distance interval to the unit length adopted.

Mathematical Representation of Time on Coordinate Axis. Let us consider a very simple example of a particle motion along the X–axis of a Cartesian coordinate system. This motion can be represented through a distance – time curve or trace on an X–T coordinate plane. The velocity and acceleration of the particle at any point along the X–axis will be represented by the slope and curvature of the trace at that point. Let us consider a particle moving in a circular orbit in XY plane. The motion of this particle can be represented as a helical trace in a XY–T coordinate space or manifold. The velocity and acceleration characteristics of this particle will be represented by the geometry of helical trace in the XY–T manifold. An important point to be noted here is that *the helical trace does not physically exist anywhere at any time*; it is just a mathematical or graphical representation of the motion of a particle over a period of time.

Similarly the motion of various particles in three-dimensional physical space XYZ can be represented through suitable traces in a four dimensional XYZ–T space-time manifold. An important point to be noted here too is that four dimensional traces of particles do not physically exist anywhere at any time; these are just mathematical representations of the motion of particles in three dimensional space over a period of time. In the same way, a four-dimensional space–time manifold XYZ–T does not physically exist anywhere; it is just a mathematical notion. However, due to the dominance of mathematics during the 20th century, the mathematical notion of space–time manifold has been assigned a more sophisticated identity of spacetime continuum. In Relativity, the notion of spacetime continuum has been treated as a physical entity, which could even be deformed and curved!! As discussed above, this is fundamentally incompatible with the basic notion of Time.

4

Absolute Reference Frame

4.1 Valid Coordinate Reference Frames

A reference frame is a set of space coordinates, fixed in some defined way. Let us consider a closed volume V of space containing a system of N particles of matter in all possible physical states. We consider a closed volume of space in the sense that there is no transfer of mass or energy across the boundary surface of this volume; and the enclosed particles do not experience any significant force or interaction from outside this volume. Let point A be the center of mass (CoM) of these N particles and let K be a non-rotating Cartesian coordinate reference frame with its origin located at point A. In this reference frame K, let the positions of all N particles be defined to be certain function of time ($x_i(t)$, $y_i(t)$, $z_i(t)$), provided they remain bounded within the closed volume V. Since K is a reference frame with origin at the center of mass of the enclosed N particles, it is generally referred as a CoM reference frame. In a CoM reference frame, the total momentum of all of its domain particles is zero.

Within the closed volume V under consideration, total momentum and total mass-energy content of the given N particles will be conserved. We may refer this set of N particles to any coordinate reference frame, for quantifying or assigning certain measure numbers to the relative positions of these particles. But that must not alter the physical state or content of matter (e.g. mass-energy content) within the closed volume V under consideration. This requirement may be treated as a physical constraint on the choice of valid coordinate reference frames.

Out of all other inertial reference frames which could be constructed for referring the positions and velocities of given N particles within a closed volume V, the total mass-energy content measured in a CoM reference frame is the minimum. Hence, a CoM reference frame may be considered as an absolute or fixed reference frame for the given N particles contained within a closed volume V. *This is the fundamental notion of an absolute reference frame in relation to matter contained within a closed volume of space.* Since the domain particles of the reference frame K do not experience any significant force or interaction from outside its domain volume, the center of mass and hence the origin A of reference frame K will continue to remain in its state of rest or of uniform motion in the external space outside its domain volume. Hence the reference frame K can also be regarded as a unique fixed Inertial Reference frame for the closed volume under consideration.

4.2 International Celestial Reference System

As a consequence of the IAU 2000 resolutions,[3] the old celestial dynamical reference system, materialized by the FK5, is replaced by the International Celestial Reference System (ICRS), which consists of the Barycentric Celestial Reference Frame (BCRF) and the Geocentric Celestial Reference Frame (GCRF), both kinematically defined by the position of same extragalactic radio sources. The origin of space coordinates defining BCRF is located at the barycenter or the CoM of our solar system. The origin of space coordinates defining GCRF is located at the geocenter or the CoM of the Earth system. The new system is kinematic, because its coordinate directions are defined through the positions of extragalactic objects, whose proper motions are assumed to be negligible in comparison with the accuracy of observations.

For making a physical comparison between the BCRF and GCRF, let us introduce an unconventional term – the physical domain volume of a reference system. Roughly speaking, we are already aware of the fact that BCRF is a much bigger reference system covering the entire solar system, whereas the GCRF is comparatively a smaller reference system covering the earth-moon system. The physical domain volume of a reference system may be defined as the volume of space which contains the locations of all material particles that are used for computing the center of mass (origin of reference frame) of that system. Now, the physical domain volume of BCRF can be defined as the volume of the whole solar system within which all material particles that are co-moving with the solar system, are located. Similarly, the domain volume of GCRF can be defined as the volume of the earth-moon system within which all particles that are co-moving with the earth system, are located. Here BCRF can be regarded as an absolute or fixed reference frame in relation to the solar system whereas the GCRF, being a subset of BCRF, can be regarded as a local reference frame in relation to the solar system.

While describing the motions of terrestrial space flights, artificial satellites or the Moon, one must use the GCRF. This natural reference system moves with the Earth around the Sun. However, for describing the motions of planets, comets and inter-planetary space missions, one must use the BCRF. The task of establishing and maintaining the ICRS and its components has been assigned to the International Earth Rotation and Reference Systems Service (IERS). Major components of IERS include Technique Centers, Product Centers and Combination Centers. The main contributing observational techniques used are, International GNSS Service (IGS), International Laser Ranging Service (ILRS), International VLBI Service (IVS) and International DORIS Service.

4.3 Critical Observations on Relativity Principle

As per the Relativity Principle: "If a system of coordinates K is chosen so that, in relation to it, physical laws hold good in their simplest form, the same laws hold good in relation to any other system of coordinates K' moving in uniform translation relatively to K." All non-rotating reference frames that move with uniform velocity with respect to one another, are defined as Inertial Reference Frames. The origins of all inertial reference frames will therefore move in straight lines. All inertial reference frames constitute a group and no particular member of this group can be considered a preferred reference frame.

Basically all laws of Nature will remain valid and operative independent of reference frames. However, in physics we quantify the laws of Nature, so as to represent them through certain mathematical equations involving dimensional physical parameters. We need the structure of coordinate systems and reference frames to quantify the physical parameters of relative positions, velocities, accelerations, force, momentum and kinetic energy of various interacting particles or groups of particles. To ensure that the laws of physics remain independent of the reference frame, the form or content of the mathematical equation representing any law of physics must not change with any change in the reference frame. Obviously therefore, some constraints will be required to be imposed on the choice of valid reference frames.

However, physical parameters of velocity, momentum and kinetic energy are not invariant in the inertial reference frames (IRF) in relative uniform motion. As such, some of the laws of physics, the representative equations of which include the parameters of velocity, momentum or kinetic energy, will no longer remain invariant in the inertial reference frames in relative uniform motion. Hence it is wrong to assume that all laws of physics are invariant in the group of inertial reference frames.

Whereas the principle of relativity gives the impression that infinitely many inertial reference frames are available to the user for use as per convenience; the elaborate arrangements required for establishing just one reference frame, the BCRF, must be a bit perplexing. Probably the notion of inertial reference frames, in relative uniform motion, is too simplistic, vague and misconstrued. Let us examine this notion critically:

- ***Why should reference frames be required to move at all?*** Logically it is the particles of matter that are expected to move in a reference frame. Primarily the reference frames are required for quantifying the positions of various particles located in a given region of space. A reference frame with its origin fixed at the CoM of all the particles in the given region of space, is sufficient to quantify the positions of all such particles. We just don't need a large number of reference frames in relative uniform motion, to quantify the positions of a given set of particles. It would be utterly illogical and misleading if the IERS created 10 more celestial reference frames in relative uniform motion with respect to the BCRF.

- ***Why do we need very many reference frames?*** For studying the kinematic motion and dynamic interactions of an infinitely large number of particles located in a given region of space (of a closed volume V), we need to reference their positions to a single CoM reference frame (like BCRF for the solar system). If we create a separate reference frame for each particle, the very objective of creating a reference frame will be lost. However, some local reference frames (like GCRF in the solar system) could always be created for the convenience of practical measurements of positions and velocities, provided such local measurements could ultimately be transformed to the fixed CoM reference frame.
- ***Can multiple IRF in relative motion be established in BCRF?*** As per the Relativity Principle all non-rotating reference frames that move with uniform velocity with respect to one another, are defined as Inertial Reference Frames. Let us consider three space ships S_1, S_2, and S_3, moving within our solar system with *relative* uniform velocity with respect to one another. Further, let us associate reference frames K_1, K_2, and K_3 with these space ships so that these reference frames also move with *relative uniform velocity* with respect to one another. Therefore, in accordance with relativity principle, these reference frames K_1, K_2, and K_3 will be defined as inertial reference frames. But apart from *relative uniform velocity* between S_1S_2, S_2S_3, S_1S_3, all three space ships S_1, S_2, S_3, could also be moving under common gravitational acceleration in BCRF. Hence we find that inertial reference frames defined as per relativity principle could actually be moving under accelerated motion in a CoM or fixed reference frame. As such the very notion of inertial reference frames under uniform relative motion is ambiguous, impractical and misleading. Apparently this notion was introduced just for conducting hypothetical thought experiments. However, such hypothetical IRF in relative uniform motion can never be practically established within BCRF.
- ***Why do we need fictitious observers on each IRF?*** Actually the notion of fictitious observers is as ambiguous and misleading as the notion of IRF in relative uniform motion. Modern advancements in technology have replaced the notion of fictitious observers with

advanced electronic instrumentation while real observers watch a computer display to observe the process. For example the position and velocity measurements of a spacecraft are first recorded in the local reference frame of instrumentation and then transformed to the CoM fixed frame of the solar system, the BCRF.

- ***Can relative measurements alone yield correct information?*** No, relative measurements alone cannot yield true information regarding position and velocity of particles in the relevant region of space under consideration. To illustrate this point, let us consider two space ships S_1 and S_2 moving in the solar system. Let their position vectors in BCRF be $\mathbf{R_1}$, $\mathbf{R_2}$ and their velocity vectors be $\mathbf{V_1}$, $\mathbf{V_2}$ respectively. The dynamic motion of these space ships will obviously be governed by the parameters $\mathbf{R_1}$, $\mathbf{R_2}$ and $\mathbf{V_1}$, $\mathbf{V_2}$. The relative separation between S_1 and S_2 will be given by $\mathbf{R_{12}} = \mathbf{R_2} - \mathbf{R_1}$ and their relative velocity will be given by $\mathbf{V_{12}} = \mathbf{V_2} - \mathbf{V_1}$. If we use only relative coordinates and measure only the relative parameters $\mathbf{R_{12}}$ and $\mathbf{V_{12}}$ (without using BCRF) we find that the dynamic motion of the two space ships is not governed by the relative parameters $\mathbf{R_{12}}$ and $\mathbf{V_{12}}$. Hence it is quite obvious that the relative measurements alone do not provide complete information as required.

4.4 Relative Measurements

Let us now elaborate some relevant aspects of the relative measurements with or without the use of IRF. The term 'relative measurement' of object B with respect to a reference frame K_1 implies the measurement of position and velocity of B relative to the origin A_1 of reference frame K_1. There are two special cases of these relative measurements depending on the state of the origin of reference frame K_1.

- When the position and velocity of the origin A_1 of reference frame K_1 are known with respect to the relevant CoM reference frame, then the relative measurements in K_1 can be regarded as local measurements, with K_1 known as a local reference frame. Such local measurements constitute a necessary step in establishing absolute measurements in the relevant CoM fixed reference frame such as the BCRF. For example, the relative measurement of position and velocity of Pioneer type spacecraft from the deep space network (DSN) stations constitute such a local measurement.

- When the position and velocity of the origin A_1 of reference frame K_1 are not known with respect to the relevant CoM reference frame, then all measurements in K_1 can be regarded as purely relative measurements, with K_1 known as a relative reference frame. If the origins A_1 and A_2 of two such relative reference frames K_1 and K_2 are known to be moving with a uniform relative velocity with respect to each other, then these relative reference frames will be known as Inertial Reference Frames of Special Relativity fame. As per the relativity principle, all IRF constitute a group and no particular member of this group can be considered a preferred reference frame. However, since a CoM fixed reference frame like BCRF can be considered a preferred reference frame for the relevant region of space, it cannot be regarded as a member of the IRF group. Hence it can be easily seen that a group of IRF can neither be practically defined, nor be established in physical space, nor be used for real practical measurements. Such a group of inertial reference frames is only a hypothetical construct used for conducting equally hypothetical 'thought experiments'.

Finally we may conclude that a CoM reference frame may be considered as an absolute or fixed or the **preferred reference frame** for the given N particles contained within a closed volume of space. The measurements in a convenient local reference frame constitute a necessary step for establishing the absolute measurements in a relevant CoM fixed reference frame. Relative measurements alone, without reference to a CoM fixed reference frame can give misleading results. For example, relative measurement of position and velocity of a uniformly moving spacecraft, from the DSN stations may indicate as if the spacecraft is periodically accelerating towards or away from the DSN stations, which is highly misleading. Purely relative reference frames, known as inertial reference frames in SR, are only useful for conducting hypothetical thought experiments and constitute a practically redundant notion.

5

Detection of the Universal Reference Frame

5.1 Notion of Universal Reference Frame

The Universal or an *absolute reference frame* may be defined as a non-rotating inertial reference frame with its origin fixed with respect to the Center of Mass (CoM) of the Universe. We know that the origin of International Celestial Reference Frame (ICRF or BCRF) is fixed at the barycenter or the CoM of the solar system. If we could locate a point O in BCRF such that O is fixed with respect to the CoM of our Universe, then a celestial reference frame with its origin at O could be identified with the Universal Reference Frame. For this, we need to determine the velocity of O in BCRF which will lead us to determine the velocity of BCRF in the Universal Reference Frame (URF). For establishing the Universal Reference Frame with reference to BCRF, we don't need to establish the location of the center of mass of the Universe. The speed of light is an isotropic constant c and the measures of distance and time are absolute in this frame. We may exploit this property to establish URF.

The MMX type interferometer experiments are based on the implied assumption that during emission or reflection of light photons their frequency or energy content is not influenced by the state of motion of the emitters or reflectors. In view of the uncertain nature of such implied assumptions, it is not valid to use interference phenomenon to detect absolute motion in the universal reference frame. Instead we must directly measure one way signal propagation times to detect absolute motion.

Relevant Technologies. For extremely fine resolutions in time measurements, modern advancements in technology relevant for the detection of Universal reference frame are listed below.

- Cesium atomic clocks (NIST-F1) with high accuracy and precision of a fraction of a nano-second. Optical atomic clocks may also be available in near future.
- Modern telecommunication engineering with accurate real time transmission and reception of high-speed data through advanced modulation techniques in the Giga-hertz carrier frequency range.
- The GPS technology has enabled high precision measurement of time and distance. GPS satellites broadcast a timing signal along with information identifying the time for which the tick corresponds. Through current GPS time transfer techniques, it is possible to achieve timing uncertainties well below one nano-second.

5.2 Experimental Set up

Let us consider a non-rotating Cartesian coordinate reference frame XYZ with origin at point O and fixed with respect to the Universal reference frame (Figure 5.1). Further, let us assume that a space ship A is moving at a uniform velocity **U** with respect to the reference frame XYZ. Let us consider another reference coordinate frame X'Y'Z' with origin at point A and moving with the space ship. That means X'Y'Z' is a local reference coordinate frame of the space ship A and is moving at velocity **U** with respect to XYZ. For convenience we may consider the space ship A to be our familiar International Space Station (ISS), the precise position of which is well known and well established in the Barycentric Celestial Reference Frame (BCRF).

Let the reference coordinate frame X'Y'Z' be oriented parallel to XYZ such that the coordinate axes AX', AY', AZ' are parallel to OX, OY, OZ respectively. In practice however, we can choose the orientation of X'Y'Z' as per our convenience and then define the orientation of coordinate frame XYZ to be parallel to it. In order to experimentally determine the velocity **U** of this space ship or the Observer Station A, a space probe B_1 is sent out in X' direction to a distance of a few thousand kilometers. Let U_1 be the component of **U** in the direction AB_1. We assume that both of the Observer Station A as well as the probe B_1, are equipped with precisely synchronized Cesium atomic clocks and identical microprocessor controlled Transponders, to transmit and receive coded signal pulses automatically. We can safely assume that at the time of commencement of the experiment, the point O coincides with A and axis OX is parallel to AB_1.

At a certain instant of time t_n let the space probe B_1 be at a distance D_n from A and moving at a uniform velocity V_1 along AB_1 with respect to reference frame XYZ, as shown in figure 5.1. Here it is not important to assume the joining line AB_1 to be always parallel to OX or AX'. But it is important to assume that the velocity U_1 is the component of velocity **U** along the joining line AB_1. Also, it is important to assume that the velocity V_1 is the component of velocity **V** of the probe B_1 along the joining line AB_1. Direction AB_1 is known in the local coordinate reference frame X'Y'Z'. Exact position of A in the reference frame XYZ is not required for this experiment. The whole experiment is conducted on the basic presumption that the speed of light and EM waves is a universal constant c in the Universal Reference Frame and hence in the XYZ frame.

Figure 5.1

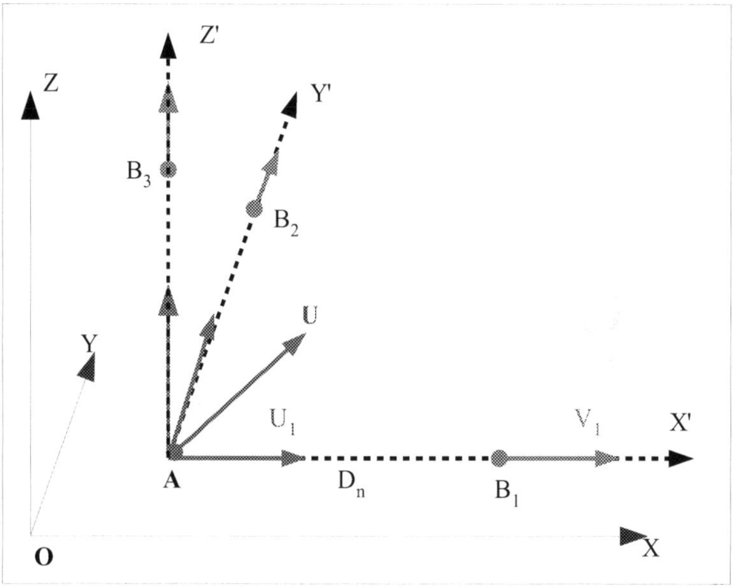

5.3 The Experimental Technique

Let us assume that at time t_n the transponder at A is triggered to send a signal pulse from A towards B_1 with coded information of t_n contained in the pulse. Let T_{da} and T_{db} be the hardware time delays in generation and transmission of the pulses at A and B_1 respectively. Thus the first pulse will actually leave transponder A at time $\tau_n = t_n + T_{da}$. Let this pulse reach B_1 at time t_{n+1} to trigger a return pulse from B_1 towards A with the coded information of time t_{n+1} contained in the return pulse. The return pulse will actually leave transponder B_1 at time $\tau_{n+1} = t_{n+1} + T_{db}$. Let this return pulse reach A at time t_{n+2} to trigger another forward pulse from A towards B_1 with the coded information of time t_{n+2} contained in the forward pulse. This forward pulse will actually leave transponder A at time $\tau_{n+2} = t_{n+2} + T_{da}$. Let this forward pulse reach B_1 at time t_{n+3} to trigger another return pulse from B_1 towards A with the coded information of time t_{n+3} contained in this return pulse; and so on.

Let this process of forward and return pulses with enclosed coded information about their times of arrival, continue for a certain period of time. This data about the time of arrival of each of the forward and return pulses will keep getting analyzed in real time in the dedicated microprocessors to compute U_1, V_1 and D_n as per the following relations. The difference in the time of origin of a new pulse and the time of arrival of previous pulse in each transponder is assumed constant, equal to the hardware time delays T_{da} and T_{db} . The hardware time delays could be determined with high precision in real time.

During the time interval t_n to t_{n+1}, out of which the signal propagation time is only $[t_{n+1} - \tau_n]$ or $[t_{n+1} - (t_n + T_{da})]$, the distance traveled by the first forward pulse is given by,

$$c. [t_{n+1} - (t_n + T_{da})] = [D_n + (V_1 - U_1).T_{da}] + V_1. [t_{n+1} - (t_n + T_{da})]$$

$$\text{or,} \quad (c - V_1). (t_{n+1} - t_n) = D_n + (c - U_1).T_{da} \qquad (5.1)$$

Here, the term $[D_n + (V_1 - U_1).T_{da}]$ represented the separation distance between A and B_1 at the instant of time τ_n when the leading edge of the signal pulse leaves the station A. At the instant t_{n+1} the leading edge of the signal pulse reaches transponder B_1. The term $V_1. [t_{n+1} - (t_n + T_{da})]$ represents the distance traveled by probe B_1 along AB_1 in the XYZ reference frame, during the propagation time of the signal pulse. During the time interval t_{n+1} to t_{n+2}, out of which the signal propagation time is only $[t_{n+2} - \tau_{n+1}]$ or $[t_{n+2} - (t_{n+1} + T_{db})]$, the distance traveled by first return pulse is given by,

$$c. [t_{n+2} - (t_{n+1} + T_{db})] = [D_{n+1} + (V_1 - U_1).T_{db}] - U_1. [t_{n+2} - (t_{n+1} + T_{db})]$$

$$\text{or,} \quad (c + U_1). (t_{n+2} - t_{n+1}) = D_{n+1} + (c + V_1).T_{db} \qquad (5.2)$$

Here too the term $[D_{n+1} + (V_1 - U_1).T_{db}]$ represented the separation distance between B_1 and A at the instant of time τ_{n+1} when the leading edge of the pulse leaves the probe B_1. At the instant t_{n+2} the leading edge of the pulse reaches the transponder A. The term $U_1. [t_{n+2} - (t_{n+1} + T_{db})]$ represents the distance traveled by station A along AB_1 during the propagation time of the signal pulse.

During the time interval t_{n+2} to t_{n+3}, out of which the signal propagation time is only $[t_{n+3} - \tau_{n+2}]$ or $[t_{n+3} - (t_{n+2} + T_{da})]$, the distance traveled by the second forward pulse is given by,

$$c. [t_{n+3} - (t_{n+2} + T_{da})] = [D_{n+2} + (V_1 - U_1).T_{da}] + V_1. [t_{n+3} - (t_{n+2} + T_{da})]$$

$$\text{or,} \quad (c - V_1). (t_{n+3} - t_{n+2}) = D_{n+2} + (c - U_1).T_{da} \qquad (5.3)$$

Since the net separation velocity between A and B_1 is $(V_1 - U_1)$, the increase in D_n during the time interval t_n to t_{n+2} is also given by,

$$(V_1 - U_1).\,(t_{n+2} - t_n) = D_{n+2} - D_n \tag{5.4}$$

Subtracting equation (5.1) from (5.3), we get

$$(c - V_1).(t_{n+3} - t_{n+2} - t_{n+1} + t_n) = D_{n+2} - D_n \tag{5.5}$$

Therefore, from equations (5.4) and (5.5) above,

$$(V_1 - U_1).(t_{n+2} - t_n) = (c - V_1).\,(t_{n+3} - t_{n+2} - t_{n+1} + t_n) \tag{5.6}$$

or, $$U_1.(t_{n+2} - t_n) = V_1.(t_{n+3} - t_{n+1}) - c.(t_{n+3} - t_{n+1}) + c.\,(t_{n+2} - t_n)$$

or, $$U_1 = V_1.(t_{n+3} - t_{n+1})/(t_{n+2} - t_n) + c - c.(t_{n+3} - t_{n+1})/(t_{n+2} - t_n) \tag{5.7}$$

Further, the increase in separation distance D_{n+1} during the time interval t_{n+1} to t_{n+2} is also given by,

$$(V_1 - U_1).\,(t_{n+2} - t_{n+1}) = D_{n+2} - D_{n+1} \tag{5.8}$$

Substituting from (5.2) and (5.3) into (5.8),

$$(V_1 - U_1).(t_{n+2} - t_{n+1}) = (c - V_1).\,(t_{n+3} - t_{n+2}) - (c + U_1).(t_{n+2} - t_{n+1}) + c.(T_{db} - T_{da}) + (V_1.T_{db} + U_1.T_{da})$$

or $$V_1.(t_{n+2} - t_{n+1}) = (c - V_1).(t_{n+3} - t_{n+2}) - c.(t_{n+2} - t_{n+1}) + c.(T_{db} - T_{da}) + (V_1.T_{db} + U_1.T_{da})$$

or $$V_1.(t_{n+3} - t_{n+1} - T_{db}) = c.(t_{n+3} - 2t_{n+2} + t_{n+1} + T_{db} - T_{da}) + U_1.T_{da}$$

or $$V_1 = c.[(t_{n+3} - 2t_{n+2} + t_{n+1} + T_{db} - T_{da})/(t_{n+3} - t_{n+1} - T_{db})] + U_1.T_{da}/(t_{n+3} - t_{n+1} - T_{db}) \tag{5.9}$$

Substituting this value of V_1 from equation (5.9) to (5.7) we get,

$$U_1 = \frac{\left[c.\left(t_{n+3} - 2t_{n+2} + t_{n+1} + T_{db} - T_{da}\right) + U_1.T_{da}\right]}{\left(t_{n+3} - t_{n+1} - T_{db}\right)} \cdot \frac{\left(t_{n+3} - t_{n+1}\right)}{\left(t_{n+2} - t_n\right)} + c.\frac{\left(t_{n+2} - t_n - t_{n+3} + t_{n+1}\right)}{\left(t_{n+2} - t_n\right)}$$

Or,

$$U_1.\left[1 - \frac{T_{da}\left(t_{n+3} - t_{n+1}\right)}{\left(t_{n+3} - t_{n+1} - T_{db}\right)\left(t_{n+2} - t_n\right)}\right]$$
$$= c.\frac{\left(t_{n+3} - 2t_{n+2} + t_{n+1} + T_{db} - T_{da}\right)}{\left(t_{n+3} - t_{n+1} - T_{db}\right)} \cdot \frac{\left(t_{n+3} - t_{n+1}\right)}{\left(t_{n+2} - t_n\right)} + c.\frac{\left(t_{n+1} - t_n\right) - \left(t_{n+3} - t_{n+2}\right)}{\left(t_{n+2} - t_n\right)}$$

Or,

$$U_1.\left[\frac{(t_{n+3}-t_{n+1})(t_{n+2}-t_n)-T_{da}(t_{n+3}-t_{n+1})-T_{db}(t_{n+2}-t_n)}{(t_{n+3}-t_{n+1}-T_{db})(t_{n+2}-t_n)}\right]$$
$$=c.\frac{[(t_{n+3}-t_{n+2})-(t_{n+2}-t_{n+1})+(T_{db}-T_{da})](t_{n+3}-t_{n+1})}{(t_{n+3}-t_{n+1}-T_{db})(t_{n+2}-t_n)}$$
$$+c.\frac{(t_{n+3}-t_{n+1}-T_{db})[(t_{n+1}-t_n)-(t_{n+3}-t_{n+2})]}{(t_{n+3}-t_{n+1}-T_{db})(t_{n+2}-t_n)}$$

Or,

$$U_1=c.\frac{(t_{n+3}-t_{n+1}).[(t_{n+1}-t_n)-(t_{n+2}-t_{n+1})]}{(t_{n+3}-t_{n+1}).(t_{n+2}-t_n)-T_{da}.(t_{n+3}-t_{n+1})-T_{db}(t_{n+2}-t_n)}$$
$$+c.\frac{T_{db}.[(t_{n+3}-t_{n+2})-(t_{n+1}-t_n)]+(T_{db}-T_{da})(t_{n+3}-t_{n+1})}{(t_{n+3}-t_{n+1}).(t_{n+2}-t_n)-T_{da}.(t_{n+3}-t_{n+1})-T_{db}(t_{n+2}-t_n)} \quad \ldots \quad (5.10)$$

Using this value of U_1 we can compute V_1 from equation (5.9). Now, using these values of U_1 and V_1, we can easily compute D_n, D_{n+1} and D_{n+2} from equations (5.1), (5.2) and (5.3) respectively.

Special Case when $V_1=U_1$: In this case when $V_1=U_1$ or the separation distance AB_1 is fixed, we can compute U_1 from equations (5.1) and (5.2) only as,

$$(c+U_1).\ [t_{n+2}-(t_{n+1}+T_{db})]=(c-U_1).\ [t_{n+1}-(t_n+T_{da})]$$

Or,

$$U_1.[t_{n+2}-(t_{n+1}+T_{db})+t_{n+1}-(t_n+T_{da})]=c.[t_{n+1}-(t_n+T_{da})-t_{n+2}+(t_{n+1}+T_{db})]$$

Or,

$$U_1.\ [(t_{n+2}-t_n)-T_{da}-T_{db}]=c.[(t_{n+1}-t_n)-(t_{n+2}-t_{n+1})+(T_{db}-T_{da})]$$

Or,

$$U_1=c.\frac{[(t_{n+1}-t_n)-(t_{n+2}-t_{n+1})+(T_{db}-T_{da})]}{[(t_{n+2}-t_n)-T_{da}-T_{db}]} \quad (5.11)$$

5.4 The Result

Hence the most important result of this analysis is the computation of the velocity component U_1 as given by equation (5.10) or (5.11). This is the component of velocity **U** of the space ship A, in the direction of AB_1. Essentially, the velocity component U_1 is found to be c times the ratio of the difference in propagation times of forward and return pulses to the sum of their propagation times.

The computation of velocity component U_1 in the direction AB_1 as shown above, is just one step towards the final determination of the velocity vector **U** of the space ship A in the XYZ coordinate reference frame. Let us further assume that space probes B_2 and B_3, similar to probe B_1, are sent out in the directions of AB_2 and AB_3 as shown in the figure 5.1. Through the same procedure as described above, we can compute the velocity components U_2 and U_3 in the directions of AB_2 and AB_3 respectively. Once we determine the three velocity components U_1, U_2 and U_3 along three non-coplanar known directions AB_1, AB_2 and AB_3, we can then easily compute the resultant velocity vector **U** of the Observer Station with respect to the XYZ reference frame. If we choose the International Space Station to be our observer station A, then three communication satellites in geostationary orbits can be chosen to be the required space probes B_1, B_2 and B_3 for this experiment. Let us assume that the direction cosines of AB_1, AB_2 and AB_3 in the local coordinate reference frame X'Y'Z' (as well as in XYZ) are known to be (l_1, m_1, n_1), (l_2, m_2, n_2) and (l_3, m_3, n_3) respectively. Therefore, the components of the resultant velocity vector **U** along the X, Y and Z axes are given by,

$$U_x = l_1 U_1 + l_2 U_2 + l_3 U_3 \tag{5.12}$$

$$U_y = m_1 U_1 + m_2 U_2 + m_3 U_3 \tag{5.13}$$

$$U_z = n_1 U_1 + n_2 U_2 + n_3 U_3 \tag{5.14}$$

Tthe magnitude of the resultant velocity **U** is therefore given by:

$$U = \sqrt{U_x^2 + U_y^2 + U_z^2} \tag{5.15}$$

From equation (5.11), it is clear that the ratio U/c depends on the ratio of the difference between the up-link and down-link signal propagation times to the total round trip signal propagation time. The minimum difference between the up-link and down-link timings depends on the precision and accuracy of the synchronized time clocks used in the experiment. For U/c of the order of 10^{-8}, time transfer accuracy of 10^{-10} seconds, the round trip signal propagation time is required to be of the order of 10^{-2} seconds that corresponds to D of the order of a few thousand kilometers. But for U/c of the order of 10^{-7}, time transfer accuracy of 10^{-12} seconds, the round trip signal propagation time is required to be of the order of 10^{-5} seconds that corresponds to D of the order of a few thousand meters.

With well below a nano-second time transfer accuracy feasible with modern technology, the net velocity **U** of the space ship could be determined to an accuracy of a few meters per second provided the separation distance D_n is of the order of a few thousand kilometers. This in turn amounts to the precision with which we can detect and establish the Universal reference frame. The computation of velocity components U_1, U_2 and U_3 mainly depends on the isotropic constant speed of light c in the chosen reference frame XYZ and a sequence of discrete precision time measurements t_n with a set of properly calibrated and synchronized atomic clocks. Essentially, the measurement of a sequence of discrete precision time intervals employs a two way time transfer technique made feasible by modern cutting-edge technology.

This leads us to the conclusion that we can experimentally determine the velocity vector **U** of the observer station A with respect to the universal coordinate reference frame XYZ. It obviously implies that with respect to our known position A, or with respect to our local coordinate frame X'Y'Z', we have determined the velocity (-**U**) of XYZ or the Universal reference frame. With this, we confirm the detection and establishment of the Universal Reference Frame.

Concluding Remarks. For conducting this experiment, we can actually use the International Space Station (ISS) as our main observer station A and three communication satellites in geostationary orbits as the space probes. With the use of currently available front line technology and special dedicated software, a continuous record of velocity **U** of the observer station ISS with respect to the Universal Reference Frame can be obtained in real time. Further since the velocity of ISS is already known in BCRF, we can now compute the velocity **U'** of the BCRF or the barycenter of the solar system in the Universal Reference Frame. Possibly, the International Earth Rotation and Reference Systems Service (IERS) could be the most appropriate agency for undertaking this project to establish and maintain the Universal Reference Frame. However, with the development of optical atomic clocks underway, the time transfer accuracy of the order of picoseconds (10^{-12} s) will be feasible. With this accuracy, a LIGO type experimental set up can be established on the ground for determining **U** with an accuracy better than 0.1 km/second.

6

Invalidity of the Special Theory of Relativity

6.1 Special Principle of Relativity

As per the Relativity Principle, all non-rotating reference frames, in which the equations of Newtonian mechanics hold good, are defined as Inertial Reference Frames (IRF). All inertial reference frames that move with uniform velocity with respect to one another constitute a group of equivalent reference frames and no particular member of this group can be considered a preferred reference frame on any ground. Albert Einstein, in his 1905 paper 'On the Electrodynamics of Moving Bodies' compares the representations in two such frames by terming one of them as a "stationary system" and the other one as a "coordinate system in uniform motion".[4]

At the outset, let us make it clear that the terms 'coordinate systems' and 'reference frames' used here and in the subsequent discussions, specifically refer to reference systems which can be physically established and used for taking physical measurements. Throughout these discussions, we do not imply to refer to similar abstract mathematical terms like coordinate spaces, coordinate manifolds and their metric representations used in abstract mathematical analysis. Quite often, a subtle mix up between abstract mathematical notions and real physical concepts can become a source of confusion.

The principle of relativity is the main founding postulate, the main pillar of the Special Theory of Relativity (SR). According to the first postulate of SR: "If a system of coordinates K is chosen so that, in relation to it, physical laws hold good in their simplest form, the same laws hold good in relation to any other system of coordinates K' moving in uniform translation relatively to K." Since in any closed volume of space with finite matter content, the CoM reference frame is always the unique preferred reference frame, the principle of relativity cannot hold any longer. Under the second postulate the requirement of constancy of velocity of light in vacuum was changed over to the requirement of constancy of the velocity of light in each of the infinitely many inertial reference frames in relative uniform motion, by sacrificing the absolute nature of space and time.

6.2 Invalidity of the Special Principle of Relativity

Basically, all laws of Nature will remain valid and operative independent of reference frames. However, in physics we quantify the laws of Nature, so as to represent them through certain mathematical equations involving various dimensional physical parameters. If certain mathematical equation representing a law of physics is written in terms of parameters measured or defined in a particular coordinate reference frame, then we can say that the law of physics is expressed in that reference frame.

Since the domain of physics primarily involves the study of particle interactions and their associated motion in space, the physical parameters of distance, velocity, acceleration, force, momentum and kinetic energy are invariably the dominant constituents of the laws of physics. We need the structure of coordinate systems and reference frames to quantify the physical parameters of relative positions, velocities, accelerations, force, momentum and kinetic energy of various interacting particles or groups of particles. Formulation of the laws of physics in such a way that they are independent of the reference frame implies that the form of the mathematical equation representing any law of physics should not change with any change in the reference frame. Let us therefore, critically examine whether any constraints are required to be imposed on the choice of valid reference frames. If it is found that the velocity, acceleration, force, momentum or kinetic energy parameters of various interacting particles or groups of particles, changes with the change in reference coordinate frame then obviously the form of the associated mathematical equation representing a law of physics will change.

There is a well known group of coordinate reference frames, called inertial reference frames (in relative uniform motion) in which the physical parameters of acceleration and inertial force do not change with the change in reference frame. For this reason the Newton's laws of motion are found to be invariant in the whole group of inertial reference frames. However, physical parameters of velocity, momentum and kinetic energy are not invariant in the inertial reference frames in relative uniform motion. As such, some of the laws of physics, the representative equations of which include the parameters of velocity, momentum or kinetic energy, will no longer remain invariant in the inertial reference frames in relative uniform motion. Hence it is *wrong to assume* that all laws of physics are invariant in the group of inertial reference frames.

In fact the very notion of IRF is ill conceived on extremely narrow and trivial considerations of a railway carriage and the embankment. It is doubtful whether Einstein had any clear idea about the possibility of establishing a BCRF type CoM reference frame for our solar system. This is apparent from his 1916 book on Relativity,[5]

"If the principle of relativity (in the restricted sense) does not hold, then the Galilean co-ordinate systems K, K^1, K^2, etc., which are moving uniformly relative to each other, will not be equivalent for the description of natural phenomena. In this case we should be constrained to believe that natural laws are capable of being formulated in a particularly simple manner, and of course only on condition that, from amongst all possible Galilean co-ordinate systems, we should have chosen one (K^0) of a particular state of motion as our body of reference. We should then be justified (because of its merits for the description of natural phenomena) in calling this system *absolutely at rest*, and all other Galilean systems K *in motion*."

Of course it can be easily seen that the Barycentric Celestial Reference Frame (BCRF), with its origin at the CoM of the solar system, is essentially a preferred or an absolute reference frame (K^0). As already discussed in chapter 4,this reference system has been established with enormous international cooperation, with the use of most advanced 'cutting edge' technology. Further, on the grounds of conservation of total momentum and total mass-energy content within the solar system, it is not possible to establish any other equivalent (CoM) Galilean co-ordinate system K^1, K^2, etc., which is moving uniformly relative to BCRF. For studying the dynamics of planetary motion or for planning inter-planetary space missions we must use the BCRF out of necessity and not out of convenience. We cannot use any other co-ordinate system K^1, K^2, etc., moving uniformly relative to BCRF. That is, we cannot use the so called inertial reference frames in relative uniform motion for tracking any spacecraft without using BCRF. This constitutes a sufficient proof of the invalidity of the principle of relativity as per Einstein's own contention.

Essentially, the invalidity of the Principle of Relativity is justified on the following grounds:

(a) The notion of IRF in relative uniform motion is practically a redundant notion mainly used for conducting hypothetical thought experiments. It is practically impossible to uniquely establish two or more IRF in relative uniform motion, without reference to a CoM fixed absolute reference frame like BCRF or GCRF.

(b) If a relative velocity $\mathbf{v}_{ab}$ between two objects A and B is specifically required for certain analysis, it can be computed from their absolute velocities $\mathbf{v}_a$ and $\mathbf{v}_b$ in the relevant CoM fixed frame as $\mathbf{v}_{ab} = \mathbf{v}_b - \mathbf{v}_a$. But if only a relative velocity $\mathbf{v}_{ab}$ is measured in an IRF, then it is impossible to retrieve the absolute velocities $\mathbf{v}_a$ and $\mathbf{v}_b$ from $\mathbf{v}_{ab}$

alone. All measurements of position and velocity of material particles made from IRF in relative motion, (without reference to a CoM fixed reference frame) yield only relative or apparent values which cannot be used in any scientific analysis or application.

(c) All measurements of position and velocity of material particles within a closed volume of space can be referred to a single CoM fixed reference frame like BCRF. For taking measurements through sophisticated instrumentation we may use some local reference frames (like GCRF in the solar system) the motion of which is precisely known in the absolute CoM reference frame. Use of inertial reference frames in relative uniform motion, is inadequate for this purpose.

All coordinate reference frames, the origins of which are in a state of motion with respect to the origin of the CoM reference frame, cannot be considered as valid reference frames for expression of the laws of physics through appropriate mathematical equations for the following reasons:

(a) Mathematical expression for kinetic energy of all material particles, involves terms with explicit velocity dependence. That is why all laws of physics involving kinetic energy of particles, will not remain invariant when expressed in different inertial reference frames. Foremost among such laws is the law of conservation of total mass-energy content within a closed or isolated volume of space. Since total mass-energy content within a closed volume of space does not remain constant when referred to different inertial reference frames in relative uniform motion, the validity of this law itself has been curtailed in relativity. This law is now being considered applicable for conservation of mass-energy content during particle interactions in any one specific reference frame. In fact the law of conservation of total mass-energy content within a closed volume of space should have been used to test the validity of coordinate reference frames for invariant expression of the laws of physics.

(b) The laws of thermodynamics are specific manifestations of the law of conservation of mass-energy as it relates to thermodynamic processes. As per kinetic theory of gases, the kinetic energy of a molecule depends upon the temperature of the gas. And the pressure of a gas is 2/3 of the mean transitional kinetic energy of the molecule in a unit volume. Since expression for kinetic energy of all gas molecules involves terms with explicit velocity dependence, it is obvious that the kinetic energy of gas molecules will appear to be different for observers in different inertial reference frames in relative uniform motion. Therefore the temperature and pressure of any portion of the gas as well as thermal equilibrium processes will appear to be different for observers in different inertial reference frames in relative uniform motion. As such the laws of thermodynamics will appear to have different form and content for observers in different inertial reference frames in relative uniform motion.

(c) Material particles of rest mass m in the solar system will appear to be of mass $\gamma.m$ (where $\gamma=1/\sqrt{(1-v^2/c^2)}$) in an inertial reference frame in relative uniform motion at velocity v, w.r.t. the BCRF. Since as per GR, the gravitational field in a certain region of space is governed by the mass-energy content (through EFE) in the vicinity, the gravitational field in the solar system will appear to be governed by $\gamma.m$ in an inertial reference frame in relative uniform motion w.r.t. the BCRF. Hence the gravitational field within the solar system will appear to be enhanced by the γ factor in an inertial reference frame in relative uniform motion w.r.t. the BCRF. This is a clear cut proof that all laws of physics do not remain invariant in different inertial reference frames in relative uniform motion w.r.t. the BCRF.

6.3 Invalidity of the Second Postulate

The second postulate of SR depicts a fundamentally and logically wrong assumption that the speed of light in vacuum is the same constant c in all reference frames in relative uniform motion. This assumption is built in to the following relation involving space-time interval dS,

$$(dS)^2 = (dx)^2 + (dy)^2 + (dz)^2 - (ct)^2 \quad \text{..... In frame K}$$
$$= (dx')^2 + (dy')^2 + (dz')^2 - (ct')^2 \quad \text{...... In frame } K^1$$
$$= (dx'')^2 + (dy'')^2 + (dz'')^2 - (ct'')^2 \quad \text{..... In frame } K^2$$

To comply with this wrong assumption, the notion of time as an absolute measure of change has been sacrificed in SR, leading to wrong notions of relative time and consequent wrong notions of length contractions. This wrong assumption has given rise to many fundamentally absurd convictions in SR. As per SR, the time intervals dt of a standard atomic clock will be seen to be different in each of the infinitely many inertial reference frames in relative motion! However, within our solar system, we use just one common CoM fixed reference frame BCRF in which the speed of light c is a constant and the measures of time and distance are absolute. The fact is that for all local reference frames K^1, K^2, etc. within our solar system, the measure of time in actual practice is the same absolute measure t (UTC) and not the relative measure t' or t''. Hence in actual practice, with the adoption of one common standard of time measurement, the notion of relative time as well as the second postulate of SR, are already rendered null and void within our solar system.

Quoting Albert Einstein, from his 1905 paper, "*If at the point A of space there is a clock, an observer at A can determine the time values of events in the immediate proximity of A. If there is at the point B of space another clock in all respects resembling the one at A, it is possible for an observer at B to determine the time values of events in the immediate neighborhood of B. But it is not possible without further assumption to compare, in respect of time, an event at A with an event at B. We have so far defined only an 'A time' and a 'B time'. We have not defined a common 'time' for A and B, for the latter cannot be defined at all unless* ***we establish by definition that the 'time' required by light to travel from A to B equals the 'time' it requires to travel from B to A.***" This arbitrary definition of 'common time' constitutes the fundamental mistake of Einstein, which ultimately leads to the invalidation of SR.

To demonstrate this mistake, let us assume that point A and B in space represents two Pioneer type spacecrafts in the outer region of the solar system. Let the separation distance AB, as measured in BCRF, be $6x10^{12}$ m which remains constant over a period of time. Let us construct an inertial coordinate system K with its origin at A. Obviously, B will be stationary in K. Let us further assume that a spacecraft tracking station measures the velocity of A and B as $\mathbf{v_a} = \mathbf{v_b} = 3x10^5$ m/s in BCRF, along direction AB. A signal pulse transmitted from A towards B will reach B in about 20020 seconds whereas a return signal pulse transmitted from B towards A will reach A in about 19980 seconds. The uplink and down link signal propagation times can be equal only if both spacecrafts A and B are at rest in BCRF. This shows that Einstein's fundamental assumption of equating the uplink and downlink signal propagation times between A and B inherently implies that both A and B are *assumed* to be at rest in the BCRF of the solar system. Since Einstein subsequently extended his notion of *common time* between A and B, to cover all IRF in relative uniform motion within the BCRF, it obviously implies that all such IRF *in relative uniform motion are assumed to be at rest* in BCRF. This simple contradiction shatters the facade of SR.

Further, let us assume that the uplink transmitter signal carrier frequency in the above example is 2 GHz. Due to the Doppler effect, the uplink carrier frequency in the signal path AB (as measured in BCRF) will be 2.002 GHz, representing an increase of 2 MHz. Similarly, the downlink carrier frequency in the signal path BA (as measured in BCRF) will be 1.998 GHz, representing a decrease of 2 MHz. This change in carrier frequency on the signal path occurs in spite of the fact that the signal frequency received at the two receivers B and A is unchanged.

6.4 Shift in Viewpoint

Rejection of the PoR calls for a major shift in our world view of the inertial reference frames and the associated fictitious observers moving along with them. In this regard, let us review some pertinent points related to the reference frames.

- *Why do we need reference frames?*

  Primary domain of Physics includes the study of interactions, interrelations and relative movements of particles or groups of particles within a closed volume of space. For this, we need to define a coordinate reference frame to quantify or assign measure numbers to the relative positions and velocities of these particles.

- *What physical constraints need to be imposed on the choice of a valid reference frame?*

  In a closed volume of space, the enclosed particles will possess a certain temperature and pressure distribution and contain some finite amount of total mass-energy content. The choice of a valid reference frame must be constrained to ensure that the physical content, the physical attributes or properties of the enclosed particles do not get altered by a change of the reference frame.

- *Why do we need very many reference frames?*

  In any particular physical situation we do not need very many reference frames. We need only one CoM fixed absolute reference frame and one local reference frame which is well defined within the absolute reference frame.

- *Why should reference frames be in relative uniform motion?*

  No, reference frames need not be in relative uniform motion although the particles or groups of particles referred to these frames could be in relative uniform motion. Only the non-rotating reference frames with their origins at rest in BCRF can be considered as equivalent to the BCRF.

- *Why do we need to attach a reference frame with every group of particles which are co-moving in external space?*

  Such local reference frames are required to define, measure and study the relative positions of all particles within the group and for further reference to a bigger CoM fixed absolute reference frame.

- *What is the notion of an observer located on a reference frame?*

  This is a fictitious notion intended to imply the measurements taken in that particular reference frame. This can be replaced by a sensor, detector or an instrumentation set used to record the measurements.

With this paradigm shift in our conceptual viewpoint regarding coordinate reference frames, we can logically discard the principle of relativity which was the founding postulate of the special theory of relativity. Purely relative reference frames, popularly known as inertial reference frames in SR parlance, can neither be uniquely established in physical space nor can be practically utilized for real life applications. These are only useful for conducting hypothetical thought experiments and hence constitute a practically redundant notion. Secondly, as shown above, the relative velocities observed from all inertial reference frames in relative uniform motion, are inherently the apparent velocities.

Hence, in any such hypothetical inertial reference frame, the apparent speed of light cannot logically be a universal constant c. Further with the adoption of one common standard of time (UTC or TAI) the second postulate of SR is practically and logically rendered invalid. However, as shown earlier, the universal constant speed c of light propagation is an inherent property of the physical space which can be verified through experimental detection of the Universal Reference Frame. Therefore, with the collapse of both founding postulates, the Special Theory of Relativity is rendered redundant and invalid. However as discussed earlier, due to the inertial property of all forms of energy, including kinetic energy, the dynamic relations of mass and momentum still remain valid, in spite of the invalidation of the Special Theory of Relativity.

7

Representation of Continuum Strain & Stress

7.1 Basic Concepts Associated with Strain

Vector Components and Transformations. Let us consider a continuum of identifiable material points which undergoes certain deformation under the action of some physical interactions or field influences. We begin our analysis with a review of fundamental definitions of relevant basic concepts. In a three-dimensional space, let us choose three orthonormal vectors $\mathbf{a}_1$, $\mathbf{a}_2$, $\mathbf{a}_3$, as our coordinate vectors. In this case any vector **R** has the representation,

$$\mathbf{R} = R^1\mathbf{a}_1 + R^2\mathbf{a}_2 + R^3\mathbf{a}_3 \tag{7.1}$$

where (R^1, R^2, R^3) are called the components or measure numbers of **R** and the vectors $\mathbf{a}_1$, $\mathbf{a}_2$, $\mathbf{a}_3$ contain the notion of unit lengths along the coordinate lines. The corresponding physical projections of **R** along coordinate directions, will be given by the product of such contravariant components with their base vectors as $R^1{*}|a_1|$, $R^2{*}|a_2|$ and $R^3{*}|a_3|$ respectively. The matrix A in the equation $\mathbf{R'} = A\,\mathbf{R}$ can be interpreted as an operator which converts a vector **R** into another vector **R'** through a transformation of its components. **R'** may be interpreted as a deformed vector produced by the operator A. It is important to note here that the deformation of vector **R** can be brought about either through the variation of its components R^1, R^2, R^3 or through the variation of the set of base vectors $\mathbf{a}_1$, $\mathbf{a}_2$ and $\mathbf{a}_3$ by change of the coordinate system.

Invariance of Space Points. In particular, we may deal with such transformation of components R^i and base vectors $\mathbf{a}_i$ such that the vector **R** itself remains invariant. The concept of invariance of mathematical objects, called vectors and tensors, under coordinate transformations, permeates the whole structure of tensor analysis. We shall suppose that a point is an invariant. In a given reference frame, a point P is determined by a set of coordinates x^i. If the coordinate system is changed, the point P is described by a new set of coordinates y^i, but the transformation of coordinates does nothing to the point itself. A set of points, such as those forming a curve or surface, is also invariant. The curve may be described in a given coordinate system by an equation, which usually changes its form when the coordinate frames are changed, but the curve itself remains unaltered, invariant. Similarly, a triply infinite set of points, constituting a 3-D space, may also be considered invariant if an infinitesimal separation distance ds between any pair of neighboring points remains invariant under admissible coordinate transformations. The notion of invariance of the arc element ds in all admissible coordinate transformations is most crucial in the formulation and efficacy of tensor analysis.

However, it is extremely important to understand that the invariance of mathematical objects, like vectors, is only with respect to coordinate transformations. In any particular coordinate system, when we define certain vector or tensor we are free to assign any value to it. But once assigned, that value will remain invariant under all admissible coordinate systems. Of course, in any particular coordinate system, we are always free to redefine that vector or tensor or to re-assign any other value to it on physical considerations.

Metric Tensor **[g_{ij}].** Consider vector components (dy^i) determined by a pair of neighborhood points P(y) and Q(y+dy) referred to orthogonal Cartesian coordinates y^i. The square of distance between points P and Q is given by the formula of Pythagoras as,

$$(ds)^2 = (dy^1)^2 + (dy^2)^2 + (dy^3)^2 \tag{7.2}$$

Here ds is called the element of arc. A change in coordinate system from y^i to x^i given by the transformation relations:

$$y^i = y^i(x^1, x^2, x^3) \tag{7.3}$$

permits us to write the relation (2) as,

$$(ds)^2 = (g_{ij})(dx^i)(dx^j) \tag{7.4}$$

with usual summation over repeated indices i & j from 1 to 3 and where the metric tensor coefficients are given by the partial derivatives of y^i as,

$$g_{ij}(x) = \frac{\partial y^k}{\partial x^i}\frac{\partial y^k}{\partial x^j} \quad \text{sum for } k = 1 \text{ to } 3 \tag{7.5}$$

Here, the equations (7.4) and (7.5) will jointly ensure that the length of the arc element ds remains invariant with the transformation of coordinates given by (7.3). Any new coordinate system will have its corresponding metric coefficients uniquely defined through relations of the type (7.3) and (7.5). In the orthogonal coordinate systems, the value of three metric coefficients g_{11}, g_{22}, g_{33} determines the magnitude of corresponding base vectors $\mathbf{a}_1$, $\mathbf{a}_2$, $\mathbf{a}_3$ as:

$$\mathbf{a}_1.\mathbf{a}_1 = g_{11}\,; \qquad \mathbf{a}_2.\mathbf{a}_2 = g_{22}\,; \qquad \mathbf{a}_3.\mathbf{a}_3 = g_{33}\,.$$

The square of arc element is,

$$(ds)^2 = g_{11}*(dx^1)^2 + g_{22}*(dx^2)^2 + g_{33}*(dx^3)^2 \tag{7.6}$$

7.2 Deformations in Continuous Media

Let us consider a finite region of the continuum representing continuous media which undergoes certain deformations. We shall use a coordinate reference frame X for quantifying or defining the relative positions of all points in this region. Further, we shall mainly focus our attention on the deformed state of the continuum and compare it with its initial un-deformed state. To begin with let us consider a point P in the initial un-deformed state. Let (x^1, x^2, x^3) be the position coordinates and $\mathbf{r}(x^1, x^2, x^3)$ be the position vector of point P. Let Q be a point in the neighborhood of P so that the vector from P to Q written as $d\mathbf{r}$ can be represented in the form,

$$d\mathbf{r} = \mathbf{a_i}\, dx^i \qquad (7.7)$$

and the square of the arc element ds in the un-deformed state is,

$$(ds)^2 = d\mathbf{r}.d\mathbf{r} = \mathbf{a_i} \cdot \mathbf{a_j}\, dx^i\, dx^j$$

Or,

$$(ds)^2 = g_{ij}\, dx^i\, dx^j \qquad (7.8)$$

where, $\mathbf{a_i}$ are the base vectors and $g_{ij} = \mathbf{a_i} \cdot \mathbf{a_j}$ are metric coefficients in the un-deformed state of the continuum.

Now let us consider the final deformed state of the continuum. In this state let the point P from the initial un-deformed state get shifted to point P' and the point Q shifted to the point Q'. Let the position vector of point P' be termed $\mathbf{r'}$. This shift in position of neighborhood points P and Q to the positions P' and Q' is generally termed as displacement of these points and essentially constitutes the deformation of the continuum under consideration. Here, let us assume that rigid body motion (i.e. translation and rotation as a rigid body) of the continuum is not possible and all displacements of points constitute pure deformation of the continuum. In the deformed state, the vector from point P' to Q' written as $d\mathbf{r'}$ can be represented in the form,

$$d\mathbf{r'} = \mathbf{b_i}\, dx^i \qquad (7.9)$$

and the square of the arc element ds' in the deformed state is:

$$(ds')^2 = d\mathbf{r'}.d\mathbf{r'} = \mathbf{b_i} \cdot \mathbf{b_j}\, dx^i\, dx^j$$

Or,

$$(ds')^2 = h_{ij}\, dx^i\, dx^j \qquad (7.10)$$

where, $\mathbf{b_i}$ are the base vectors and $h_{ij} = \mathbf{b_i} \cdot \mathbf{b_j}$ are metric coefficients in the deformed state of the continuum.

The displacement of point P to P' is represented by a displacement vector $\mathbf{U}$ and the corresponding displacement of its neighborhood point Q to Q' is given by the incremented displacement vector $\mathbf{U}+d\mathbf{U}$. The complete deformation of the continuum can be said to be fully determined when the displacement of every point P in the continuum is known or uniquely determined. The existence of displacement vector **U** at every point P, as a function of position coordinates, will constitute a displacement vector field **U** in the continuum. The displacement vector from point P to P' is given by the relation,

$$\mathbf{U} = \mathbf{r}' - \mathbf{r}$$
$$= u^i \mathbf{a}_i \tag{7.11}$$

where u^i are the contravariant components of vector **U**. Differentiating equation (7.11) we get,

$$\partial\mathbf{U}/\partial x^i = \partial\mathbf{r}'/\partial x^i - \partial\mathbf{r}/\partial x^i = \mathbf{b}_i - \mathbf{a}_i \tag{7.12}$$

Or, $$\mathbf{b}_i = \mathbf{a}_i + \partial\mathbf{U}/\partial x^i \tag{7.13}$$

7.3 Representation of Infinitesimal Strain – Strain Tensor

Excluding rigid body motion, as already assumed, an infinitesimal deformed state of the continuum can be described as the strained state. The strained state is represented by a strain tensor E with its components e_{ij} defined at every point P of the continuum. In the linear or infinitesimal theory of deformation the strain tensor components e^i_j are computed from the covariant derivatives of the displacement vector as,

$$2\, e^i_j = u^i_{,j} + u^j_{,i} \tag{7.14}$$

and $$u^i_{,j} = \partial u^i/\partial x^j + \Gamma^i_{\alpha j}\, u^\alpha \qquad \text{(summation over } \alpha)$$

where, Γ^i_{jk} is a Christoffel symbol of second kind. However, the strained state of the continuum can also be represented by the metric h_{ij} of the deformed sate. We can say that the continuum representing the continuous media, is strained whenever arc element ds' given by equation (7.10) is different from the arc element ds given by equation (7.8). The covariant strain tensor components e_{ij} are related to this difference through following relations.[6]

$$(ds')^2 - (ds)^2 = (h_{ij} - g_{ij})\, dx^i\, dx^j$$
$$= 2\, e_{ij}\, dx^i\, dx^j \tag{7.15}$$

where, $$e_{ij} = g_{i\alpha}\, e^\alpha_j$$

and $$2\, e_{ij} = h_{ij} - g_{ij} \tag{7.16}$$
$$= \mathbf{b}_i.\mathbf{b}_j - \mathbf{a}_i.\mathbf{a}_j$$

Ideally speaking, we should be in a position to obtain the displacement vector field **U** (equation (7.11)) for the strained state of the continuum and then compute the components of the strain tensor (equation (7.14)) and the components of the modified metric (equation (7.16)). However, on physical considerations we may fix or specify the components of the strain tensor first and then work out the displacement vector field **U**. Physical constraints demand that the displacement vector field components must be finite, continuous, single valued and piecewise smooth functions of coordinates. Therefore, if we are required to compute the displacement vector field from the specified strain components, certain compatibility or integrability conditions have to be imposed on the specified strain tensor components e_{ij}. Such conditions were deduced for the linearized case by B. Saint Venant in 1860. These are known as Saint Venant's compatibility equations and are given below.

$$e_{ij,kl} + e_{kl,ij} - e_{ik,jl} - e_{jl,ik} = 0 \qquad (7.17)$$

Noting that g_{ij} are the metric coefficients of the un-deformed Euclidean space, the compatibility equations (7.17) can be shown to be derived from the fact that the Riemann tensor R^i_{jkl} based on g_{ij} vanishes.

$$R^i_{jkl} = \partial/\partial x^k\, \Gamma^i_{jl} - \partial/\partial x^l\, \Gamma^i_{jk} + \Gamma^i_{\alpha k}\Gamma^\alpha_{jl} - \Gamma^\alpha_{jk}\Gamma^i_{\alpha l} \text{ (summation over } \alpha) \qquad (7.18)$$

We can also construct Riemann tensor by using the strain tensor components e_{ij} as the metric coefficients (in the definitions of Γ^i_{jk}). Therefore, it can further be shown that if the Riemann tensor R^i_{jkl} based on e_{ij} does not vanish then the compatibility equations (7.17) cannot be satisfied. This implies that the specified strain components, with non-zero Riemann tensor, can not yield the desired displacement vector field **U** with finite, continuous, single valued and piecewise smooth components.

7.4 Representation of Infinitesimal Strain – Modified Metric

There is yet another situation where instead of specifying strain tensor components e_{ij} on physical considerations, we may specify the modified metric coefficients h_{ij} of the deformed state to represent the strained state of the continuum. From the modified metric coefficients h_{ij} we can first compute the strain tensor components e_{ij} as per equation (7.16) and then work out the displacement vector field **U** as per equation (7.14). Here too, the strain tensor components will have to satisfy the compatibility equations (7.17) to ensure proper solution of relevant partial differential equations yielding the desired displacement vector field **U** with finite, continuous, single valued and piecewise smooth components.

In actual practice however, the use of modified metric coefficients h_{ij} is limited to the development of theoretical models representing deformed state of continuous media in bulk. In practical applications this approach is rarely adopted. Let us therefore, illustrate the use of modified metric coefficients to represent the strained state of the continuum through some very simple (even trivial) examples:

- Consider a thin long string as a one dimensional continuum of identifiable material points. The string is aligned along the x^1 axis with the base vector $\mathbf{a_1}=1$ and the metric coefficient $g_{11}=\mathbf{a_1}.\mathbf{a_1}=1$. Now let us suppose that under the influence of a high temperature environment, the string gets deformed (elongated) with the modified metric coefficient h_{11} given by $(1+k)^2$. Here, k is an extremely small fraction ($k<<1$) such that $(1+k)^2$ can be written as approximately equal to $1+2k$. From equation (7.16) we get $e_{11}=k$ and modified base vector $\mathbf{b_1}=1+k$. From equation (7.14) we get $du^1/dx^1 = k$, so that the displacement vector component u^1 is given by $k.x^1$. This displacement vector completely represents the strained state of the string.
- Now consider a thin large metal sheet as a two dimensional continuum of points. The sheet is spread in x^1, x^2 plane with its edges aligned along orthogonal x^1 and x^2 axes. The metric coefficients g_{ij} are given as $g_{11} = g_{22} = 1$ and $g_{12} = g_{21} = 0$. Let us suppose that under the influence of certain high pressure environment, the sheet gets deformed (elongated) with the modified metric coefficients given by $h_{11}= h_{22}= (1+k)^2$ and $h_{12} = h_{21} = 0$. With $k<<1$, from equation (7.16) we get $e_{11} = k$, $e_{22} = k$, $e_{12} = e_{21} = 0$ and modified base vector $\mathbf{b_1}=1+k$, $\mathbf{b_2}=1+k$. From equation (7.14) we get $du^1/dx^1 = k$; $du^2/dx^2 = k$, so that the displacement vector components are $u^1 = kx^1$ and $u^2 = kx^2$. This displacement vector completely represents the strained state of the sheet.
- In our next illustrative example let us consider a case where we assume the displacement vector field **U** and then work out the modified metric coefficients that represent the strained state of the continuum. Consider a huge metal sphere of radius R (say, of the order of the radius of sun). Let point O be the center of the sphere. Let us refer the continuum of identifiable material points of the

sphere to a spherical polar coordinate system given by $x^1 = r$, $x^2 = \theta$ and $x^3 = \phi$ coordinates and with the origin at point O. The non-zero metric tensor components g_{ij} for this coordinate system are $g_{11} = 1$; $g_{22} = r^2$; $g_{33} = r^2\sin^2\theta$. Physical components u^r, u^θ, u^ϕ of displacement vector **U** are related to the corresponding contravariant components u^1, u^2, u^3 as $u^r = u^1$, $u^\theta = r\,u^2$, $u^\phi = r\sin\theta\, u^3$. Let us now assume that under gravitational influence the spherical region of the continuum gets strained such that the radial displacement vector field is given by $u^1=k.r$ and $u^2=u^3=0$ (where k is a small negative fraction such that $|k|<<1$). The corresponding non-zero strain components are given by $e^1{}_1=k$, $e^2{}_2=k$ and $e^3{}_3=k$ or $e_{11}=k$, $e_{22}=k.r^2$ and $e_{33}=k.r^2\sin^2\theta$. With these strain components we can now compute modified metric coefficients h_{ij} from equation (7.16) as, $h_{11}=g_{11}+2.e_{11} = 1+2k = (1+k)^2$; $h_{22}= g_{22}+2.e_{22} = (1+k)^2r^2$ and $h_{33}=(1+k)^2r^2\sin^2\theta$. These modified metric coefficients h_{ij} now fully represent the strained state of the spherical region of the continuum under consideration.

The foregoing analysis of the strained state of the continuum is essentially centered on the geometrical concepts, notions and definitions. There is no bearing of any interaction, forces, elasticity or dynamics involved in the said analysis. The study of motion or changes in position of all physical points of a continuum can be of great importance in understanding the physical state of that continuum. Even in the study of our familiar space continuum, it is extremely important to understand beforehand whether we intend to keep all points of the space continuum relatively fixed under all situations or do we intend to permit their relative movements under certain environments, leading to a deformed state of the space continuum.

As shown above, the notion of deformed or strained state of the continuum under study is derived from the variability or invariance of arc element ds. Whenever the arc element ds changes over to ds' under certain situations, the changed state of the continuum will be termed the deformed or strained state. The strained state can be considered fully defined or fully determined once we know or uniquely determine the displacement vector field at all points of the continuum. The strained state can also be defined through specification of strain tensor components provided these strain components satisfy Saint Venant's compatibility equations. Finally, the strained state can also be defined through specification of modified metric coefficients from which the required strain tensor components can be computed subject to the compatibility conditions. However, the compatibility conditions require that the modified metric must be Euclidean to ensure that the resulting strained state of the continuum corresponds to smooth, finite and continuous displacement components and to avoid discontinuities within the continuum. This fact is of crucial importance for examining the validity of the current mathematical model of General Relativity where the deformation of space is presented as 'curvature' of space.

7.5 Stress Tensor in an Elastic Continuum

Representation of Stress. At any point $P(x^1,x^2,x^3)$ of the Elastic Continuum under infinitesimal deformation, the state of stress is represented by stress tensor **T**, the components τ^i_j of which are defined as follows. With point $P(x^1,x^2,x^3)$ as the center, consider an infinitesimal plane rectangular surface area $\sigma_1 = \delta x^2.\delta x^3$, with its normal parallel to X^1- axis (Fig. 1). This infinitesimal area will have two faces. We shall consider that face of σ_1, where its unit normal ν_1 points towards positive X^1-axis, as +ve face and denote it as σ_{+1}.

Fig. 7.1 Representation of stress components on a surface element σ_{+1}

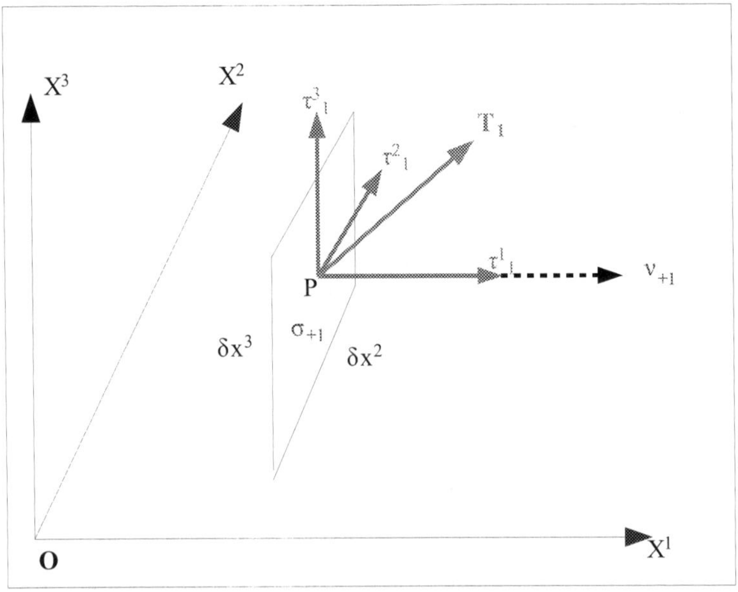

The other face, with normal pointing towards negative X^1-axis, will be considered -ve face and denoted as σ_{-1} . If the net force per unit area acting on σ_{+1} is termed $\mathbf{T_1}$, then it is obvious that the direction of $\mathbf{T_1}$ will not coincide with unit normal ν_1 in general, since this net force represents a resultant of three components. In fact the vector $\mathbf{T_1}$ acting on σ_{+1}, can be decomposed into its components along X^1, X^2 and X^3 coordinate directions (with base vectors $\mathbf{e_1}$, $\mathbf{e_2}$ and $\mathbf{e_3}$) as,

$$\mathbf{T_1} = \mathbf{e_1}\tau^1{}_1 + \mathbf{e_2}\tau^2{}_1 + \mathbf{e_3}\tau^3{}_1$$

$$= \mathbf{e_i}\tau^i{}_1 \qquad (7.19)$$

With the same point $P(x^1,x^2,x^3)$ as the center, if we now consider another plane rectangular surface area $\sigma_2 = \delta x^1.\delta x^3$, with its normal parallel to X^2-axis, the net force per unit area $\mathbf{T_2}$ acting on σ_{+2} will then be given by,

$$\mathbf{T_2} = \mathbf{e_1}\tau^1{}_2 + \mathbf{e_2}\tau^2{}_2 + \mathbf{e_3}\tau^3{}_2 = \mathbf{e_i}\tau^i{}_2 \qquad (7.20)$$

Similarly, $$\mathbf{T_3} = \mathbf{e_1}\tau^1{}_3 + \mathbf{e_2}\tau^2{}_3 + \mathbf{e_3}\tau^3{}_3 = \mathbf{e_i}\tau^i{}_3 \qquad (7.21)$$

In general, for an infinitesimal rectangular plane area σ_{+j} perpendicular to x^j coordinate direction, the net force per unit area $\mathbf{T_j}$ acting on σ_{+j} will be given by

$$\mathbf{T_j} = \mathbf{e_1}\tau^1{}_j + \mathbf{e_2}\tau^2{}_j + \mathbf{e_3}\tau^3{}_j$$

$$= \mathbf{e_i}\tau^i{}_j \qquad (7.22)$$

Here the quantities $\tau^i{}_j$ are the components of the stress tensor T at point $P(x^1,x^2,x^3)$. The stress components $\tau^i{}_j$ in general will be functions of space coordinates (x^1,x^2,x^3) of point P and time t.

The stress components $\tau^i{}_j$ are reckoned +ve if the corresponding components of force act in the directions of increasing x^i, when the surface normal is along increasing x^j axis. If on the other hand, the surface normal is along the -ve x^j axis, then positive values of components $\tau^i{}_j$ are associated with forces directed oppositely to the positive directions of x^i coordinate axes. Hence, for an infinitesimal volume element $\delta V = \delta x^1.\delta x^2.\delta x^3$ taken in the shape of a rectangular parallelepiped, with faces parallel to coordinate planes and point

$P(x^1,x^2,x^3)$ as its center, the stress components τ^i_j will correspond to forces in opposite directions at the opposite ends of the parallelepiped.

Equilibrium Equations in an Elastic Continuum. Ordinary material bodies, under stress, will generally be in a state of static equilibrium. However, in an Elastic Continuum, the equilibrium in a strained state is essentially dynamic. In a steady state or static equilibrium, not only the resultants of all forces acting on an infinitesimal volume element δV should vanish but the resultant moment of all forces should also vanish to ensure that pure stresses and strains do not give rise to rigid body motions and rotations. In the static equilibrium of a material body under stress, vanishing of resultant moments to avoid rigid body rotations can be ensured by the symmetry of stress and strain components τ^i_j and S^i_j. However, this condition is not applicable for an Elastic Continuum where there is neither static equilibrium nor rigid body rotations.

8

Fallacious Notion of Spacetime Continuum

8.1 The Notions of Space and Time

Introduction. Most followers of Relativity theories consider the spacetime continuum to be a physical entity which can even be deformed and curved. This misconception is quite deep rooted in the metaphysical eternalist viewpoint of existence in contrast to the logical presentist viewpoint. As per the eternalist viewpoint, a so-called material object in a spacetime world is a continuous series of spacetime events, each of which exists eternally as a distinct part of the world. There is no distinction between the past, present and future. We may refer to it as a block view of spacetime. As per the presentist viewpoint, the present moment is different from the past and future and that physical entities exist only in the present. The physical phenomenon does not exist in the past and the future regions of time. The foundations of General Theory of Relativity (GR) are critically dependent on the integrity of the notion of spacetime continuum. Actually, spacetime is just a mathematical notion which has no physical existence.

The Coordinate Space. The association of the set of points P on coordinate line X with the set of real numbers x, constitutes a coordinate system of the one-dimensional space, once the notion of certain unit length has been defined. The one-to-one correspondence of ordered pairs of numbers with the set of points in the plane X^1X^2 is the coordinate system of the 2D space consisting of points in the plane. Similarly, with a predefined notion of unit length, an essential feature of 3D space is the concept of one-to-one correspondence of points in space with the ordered sets of real numbers. The predefined notion of unit length or scale for different coordinate axes constitutes the metric of space for quantifying the notion of distance and the position measurements of the sets of points in this coordinate space.

We define a space (or manifold) of N dimensions as any set of objects that can be placed in a one-to-one correspondence with the ordered sets of N numbers x^1, x^2,..., x^N. Any particular one-to-one association of the points with the ordered sets of numbers is called a coordinate system and the numbers x^1, x^2, ..., x^N are termed the coordinates of points. In all coordinate spaces that are metricized, we associate the notion of unit length along all coordinate axes and a metric tensor g_{ij} with each coordinate system. All essential metric properties of a metricized space are completely determined by this tensor.

The Physical Space. The notion of physical space implies the spatial extension of the universe wherein all material particles and all fields are embedded or contained. The true void between material points is in essence the physical space, or empty space, or free space. It is important to note here that the coordinate space, along with its scale or metric, is our 'human' creation intended to facilitate the quantification of relative positions of material particles and fields. The existence of physical space does not depend in any way on the existence or non-existence of coordinate systems and coordinate spaces. Of course, for the study and analysis of physical space and the material particles and fields embedded in it, we do need the structure of coordinate systems and coordinate spaces as a quantification tool. The most significant point to be highlighted here is that whereas the metric scaling property is only associated with coordinate space, the physical properties of permittivity, permeability and intrinsic impedance are associated with physical space.

Notion of Time. In Nature, there are a large number of physical processes, which undergo cyclic changes. Depending on the consistency of such cyclic changes and the convenience of their measurement, we may select any one of them as our reference scale for relative measurement of change. The angular position of a planet in orbit, the position of a pendulum oscillating about a mean and the vibrations of many electro-mechanical systems are all examples of physical processes that undergo cyclic changes. Any such process could be adopted as a reference scale for relative measurement of change or the reference scale for time. In general, the study of natural phenomenon invariably involves the comparative study of various changes. For this comparative study, we need to use a reference scale, or more correctly a reference time scale, for relative measurement of change or for measurement of time. Hence Time, as a relative measure of change, is an important parameter in the study of an essentially dynamic physical Universe.

8.2 Particle Traces

Consider a very simple example of a particle motion along the X-coordinate. This motion can be represented through a distance-time curve or trace on an X-T coordinate plane. The velocity and acceleration of the particle at any point along the X-axis will be represented by the slope and curvature of the trace at that point. Let us now consider a particle moving in a circular orbit in XY plane. The motion of this particle can be represented as a helical trace in a XY-T coordinate space or manifold. The velocity and acceleration characteristics of this particle will be represented by the geometry of helical trace in the XY-T manifold. An important point to be noted here is that the helical trace does not physically exist anywhere at any time; it is just a mathematical or graphical representation of the motion of a particle over a period of time.

Space-time manifold XYZ-T Similarly the motion of various particles in three-dimensional physical space can be represented through suitable traces in a four-dimensional XYZ-T space-time manifold. An important point to be noted here too is that four-dimensional traces of particles do not physically exist anywhere at any time; these are just mathematical representations of the motion of particles in three dimensional space over a period of time. The 4-D geometry of particle traces is just a mathematical representation. In the same way, a four-dimensional space-time manifold XYZ-T does not physically exist anywhere at any time; it is just a mathematical notion. However, due to some logical fallacy, the mathematical notion of space-time manifold got assigned a more sophisticated name of spacetime continuum which is generally implied to be a physical entity in Relativity Theories. This notion of 'spacetime continuum' is fallacious.

The 4-D spacetime manifold can be regarded as a mathematical continuum of points (x, y, z, t). But a mathematical continuum of points cannot be said to get curved. Neither can the geometry of a 4-D mathematical continuum of points get influenced by the matter-energy content embedded in 3-D physical space. For the operation of GR, spacetime continuum of physical nature is definitely required. But for the existence of physical spacetime continuum, the physical space is required to exist at each and every point t of the time-axis. However, it can be shown that physical space can exist only at the present instant $t=t_p$ on the time-axis.

8.3 Physical Entity

The dictionary meaning of a physical entity is an entity that has physical existence. Here physical implies 'having substance or material, perceptible to the senses'. It includes quantities that can be physically measured. In essence, it implies a distinction between abstract and physical entities. If we have complete information about certain entity and can mentally visualize it, then that entity must be a physical entity (e.g. Solar System, Sound Waves etc.). If we have complete information about certain entity and still cannot mentally visualize it, then that entity must be an abstract entity (e.g. 4-D spacetime). However, if we know certain entity to be physical and still cannot visualize it then it will imply that we do not possess complete information about that entity (e.g. electron, proton etc.). The dynamic motion of particles embedded in 3-D physical space could be represented as traces in the 4-D spacetime manifold. The fact that the geometrical shape of such traces could be adjusted by manipulating the metric of this manifold, has been misconstrued to imply that the physical phenomenon of gravitation could somehow adjust the metric of the 4-D spacetime manifold. This misconception created the popular impression as if the spacetime continuum is a physical entity.

8.4 Space-time: 4-D Block or a Trace of 3-D Space?

Let us examine another crucial aspect of the notion of spacetime. That is, whether the abstract notion of spacetime manifold is a useful construct to represent a trace of 3-D physical space along the time coordinate or does it represent a 4-D block of the continuum of space and time points (x,y,z,t). In the block view of the spacetime continuum there are two crucial implications which probably have never been highlighted or critically examined:

(a) Firstly, the time coordinate is treated at par with space coordinates. That is, just as there is no a-priori bar on any two mutually interacting particles from occupying different positions on any spatial coordinate, similarly there is no a-priori bar on any two *mutually interacting particles* from occupying different positions on time coordinate (including positions in past and future time zones).

(b) Secondly, in the mathematical handling of the time coordinate, no distinction is made between the present time and the past time or the future time. That is, any material particle located at (x_1, y_1, z_1, t_1) of the spacetime continuum can influence the geometry (topography) of spacetime at locations (x_2, y_2, z_2, t_2) and (x_1, y_1, z_1, t_3) regardless of whether t_2 and t_3 represent the past time zone or the future time zone with respect to t_1.

To illustrate these points, let us consider a 2-D (thin) metal sheet located in the XY plane of a rectangular 3-D (XYZ) manifold. Let this plane 2-D sheet be positioned at $Z=z_0$ at time $T=t_0$. Let us examine the following three cases:

Case I : Traces in XYZ manifold. Let us further assume that this 2-D sheet is moving along Z-axis at a uniform velocity v with its plane surface constrained in the XY plane. Let the position of this sheet move to z_1 at time t_1, z_2 at time t_2 ... and z_n at time t_n. Suppose we wish to study the motion of free electrons constrained on the surface of this sheet and want to obtain detailed representation for their trajectories or traces of their paths over a finite period of time. For this purpose, we may find it convenient to use 3-D XYZ manifold to represent the curved traces of the particles under study. Here it can be easily appreciated that while the particles under study are constrained to move in the 2-D plane of the metal sheet, their curved traces can be represented in the 3-D XYZ manifold. It is also true that the study of the geometry of curved traces can provide us valuable information on the velocities and accelerations of the corresponding particles. A significant point to be noted here is that while an abstract XY plane can be considered as located at every point z of the Z-axis, the 2-D metal sheet (a physical entity) constrained in the XY plane can exist only at one point z_n at time t_n. That is, even though the 2-D metal sheet constrained in the XY plane does steadily traverse the Z-axis, the particles of the metal sheet cannot be said to constitute a 3-D continuum in the 3-D XYZ manifold. Further, the geometry of the 3D XYZ manifold cannot influence the velocities and accelerations of free electrons constrained on the plane surface of the metal sheet under any circumstances.

Case II : Traces in XYT manifold. Let us now assume that the 2-D plane sheet under consideration is fixed at $Z= z_0$ and does not move in any direction. Suppose, we wish to study the motion of free electrons constrained on the surface of this sheet and want to obtain detailed representation for their trajectories or traces of their paths over a finite period of time. For this purpose, we may find it convenient to use 3D XYT manifold to represent the curved traces of the particles under study. While the particles under study are constrained to move in the 2D plane of the metal sheet, their curved traces can be represented in the 3D XYT manifold. It is also true that the study of the geometry of curved traces can provide us valuable information on the velocities and accelerations of the corresponding particles. Further, the geometry of the XYT manifold cannot influence the free electrons constrained on the plane surface of the metal sheet but may influence the representation of their traces.

Let us focus on the position of the metal sheet under consideration on the time axis. Let the time axis extend from zero to infinity. Further let t_p depict the present time on the time axis. Obviously, the t_p marker is continuously moving away from the origin of the time axis. The time zone $t< t_p$ represent the past and the time zone $t> t_p$ represent the future. Now let us take a mental snapshot of the whole range of time axis. We find that the physical body of the metal sheet is only located at $t= t_p$ and is not located anywhere in the past or the future time zones. The traces of free electrons constrained to move on the surface of this sheet (physically located at $t=t_p$) can only be represented in the past time zone. However, the computed or projected trajectories of these particles can be represented in the future time zone. The most significant point to be noted from our mental snapshot of the whole range of time axis is that the physical phenomenon is occurring only in the metal sheet constrained in the XY plane and located at $t=t_p$. There is no physical phenomenon in the past or the future time zones of the XYT manifold. These past and future time zones of the XYT manifold can only be used for representing the traces or computed trajectories of the particles constrained in the present zone ($t=t_p$). Hence we can logically conclude that the past and future time zones of the XYT manifold cannot be regarded as physical entities since these are only abstract mathematical constructs. It may be emphasized here that while an abstract XY plane can be considered as located at every point t of the Time-axis, the 2-D metal sheet (a physical entity) constrained in the XY plane can exist only at one point $t= t_p$ on the Time-axis. That is, even though the 2-D metal sheet constrained in the XY plane does steadily traverse the T-axis, the particles of the metal sheet cannot be said to constitute a '3-D continuum' in the 3-D XYT manifold.

Case III : Traces in XYZT or the spacetime manifold. Let us now replace the 2-D plane sheet of case II above with 3-D physical space (e.g., the physical space associated with our solar system). Suppose we wish to study the motion of particles contained within this space and want to obtain detailed representation for their trajectories or traces of their paths over a finite period of time. For this purpose, we may find it convenient to use 4-D XYZT manifold to represent the curved traces of the particles under study. While the particles under study are constrained to move in the 3-D physical space, their curved traces can be represented in the 4-D XYZT manifold. It is also true that the study of the geometry of curved traces can provide us valuable information on the dynamics of the corresponding particles. Further, the geometry of the 4-D XYZT manifold cannot influence the dynamics of particles contained in the 3-D physical space of our solar system but may influence the representation of their traces.

To appreciate this point, we need to focus on the position of the physical space on the time axis. Let us take a mental snapshot of the whole range of time axis. We find that the physical space is only located at $t= t_p$ and is not located anywhere in the past or the future time zones. The traces of particles, constrained to move in the physical space (located at $t=t_p$), can only be represented in the past time zone. However, the computed or projected trajectories of these particles can be represented in the future time zone. The most significant point to be noted from our mental snapshot of the whole range of time axis is that the physical phenomenon is occurring only in the 3-D physical space located at $t=t_p$.

There is no physical phenomenon in the past or the future time zones of the XYZT manifold. These past and future time zones of the XYZT manifold can only be used for representing the traces or computed trajectories of the particles contained in the physical space located at the present zone ($t=t_p$). Hence, we can logically conclude that the past and future time zones of the XYZT or spacetime manifold cannot be regarded as physical entities; these are only abstract mathematical constructs. It may be emphasized here that while an abstract XYZ manifold can be considered as located at every point t of the Time-axis, the 3-D physical space constrained in the XYZ manifold can exist only at one point $t= t_p$ on the Time-axis. That is, even though the 3-D physical space constrained in the XYZ manifold does steadily traverse the T-axis, the points of the physical space cannot be said to constitute a 4-D continuum in the 4-D XYZT manifold.

Hence, the spacetime continuum is not a physical entity but just an abstract mathematical notion which can neither influence any physical phenomenon nor can its geometry be influenced by any physical phenomenon. But GR is based on the fallacious notion of a physical 4-D spacetime continuum, where the mass-energy content within a finite region of 3-D physical space (say our solar system) is required to govern the metric of 4-D spacetime manifold in the vicinity, in accordance with Einstein's Field Equations. It is unbelievable that the matter-energy content of the solar system located on the present time zone on the Time-axis, can physically influence the geometry of 4-D spacetime in its past and future time zones. Thus the 'Castle in the Air' of Relativity, founded on the fallacious notion of 'spacetime continuum' collapses as a conceptual mistake. Even if the spacetime continuum is regarded as a physical entity as per the metaphysically eternalist viewpoint, the GR can still be shown to be invalid on the grounds of discontinuous deformations induced under the GR postulate.

8.5 Basic Objectives of the Spacetime Model

The spacetime structure has been exploited for two major objectives in Relativity. The first objective is to ensure a constant speed of light propagation in all inertial reference frames moving at uniform relative velocity with respect to one another, by adopting Minkowski spacetime manifold. The second, and more prominent objective, is to employ the 4D spacetime manifold as a graphical template to facilitate the plotting of trajectories of objects moving in a gravitational field.

Spacetime Minkowski Manifold. In SR it has been assumed that speed of light in vacuum will be constant c in all inertial frames K_1, K_2, etc. in relative uniform motion. This assumption has been built into the invariance of spacetime interval as:

$$dS^2 = (dx)^2 + (dy)^2 + (dz)^2 - (ct)^2$$
$$= (dx')^2 + (dy')^2 + (dz')^2 - (ct')^2$$
$$= (dx'')^2 + (dy'')^2 + (dz'')^2 - (ct'')^2 \quad (8.1)$$

This distinguishes Special Relativity from Galilean relativity and is the origin of the special concept of spacetime due to the special linkage between space and time coordinates through equation (8.1).

As a consequence of this assumption the measure of time and distance becomes relative and different in each of the frames K_1, K_2, etc. in accordance with Lorentz transformations, and this is important. All notions of length contraction or time dilation or clock synchronization originate from the operation of equation (8.1) for different inertial frames K_1, K_2, etc. in relative uniform motion. However, within our solar system, BCRF is the unique reference system K_0 in which the speed of light c is a constant and the measure of time and distance is absolute not relative. Within the solar system, there may be an infinitely large number of particles or objects in relative motion (e.g. trains, aircraft, satellites, planets etc.) with which we could attach reference frames K_1, K_2, etc. But the measure of time is still required to be the same absolute measure UTC or TAI. Here, it is not important which coordinate frame you actually use for convenience. It is important to note that for all such convenient coordinate frames you actually use a common time measure say UTC, the usage of which has been rendered extremely convenient through GPS receivers. Usage of this common measure of time in all frames K_1, K_2, etc. has effectively rendered the notions of spacetime and relative time t' or t'' as obsolete in real life.

8.6 Spacetime manifold as a Graphical Template

For plotting certain data on ordinary graph paper with rectangular Cartesian X-Y axes, we generally use uniform scales on each axis. As is well known, by using a logarithmic scale, one can convert an exponential curve to a straight line. A log-log plot is a two-dimensional graph of numerical data that uses logarithmic scales on both the horizontal and vertical axes. Due to the nonlinear or differential scaling of the axes, a function of the form $y = a.x^b$ will appear as a straight line on a log-log graph, in which b will be the slope of the line.

Let us now consider the trajectory of an object, falling vertically on a gravitating body of mass M. We can plot this trajectory on a Y-T graph such that Y-axis represents the height and T-axis represents time.

We find this trajectory to be a parabolic curve. Now taking a cue from the log-log graph, we can choose a suitable log-log or differential scale along Y and T axes such that the parabolic trajectory on normal graph transforms into a straight line on the differential scale graph. This makes the trajectory of an object, moving in a gravitational field, look very simple. Let us use this differential scale graph as a template. For obtaining the trajectory of any other object falling vertically on a gravitating body of mass M, we only need to locate the initial starting position of this object on the template graph and then draw a straight line of the required slope.

Let us now consider the trajectory of an object, falling in 2D X-Y plane on a gravitating body of mass M. We can plot this trajectory on a 3D XY-T linear scale graph. We find this trajectory to be a complex spiraling curve. Now we can choose a suitable log-log or differential scale along X,Y and T axes such that the spiraling trajectory on linear scale graph transforms into a straight line equivalent geodesic on the differential scale graph. Once again, this makes the trajectory of an object, moving in a gravitational field, look very simple. Let us make a template of this differential scale graph. The differential scale or the magnitude of the unit vectors along any particular axis of an orthogonal coordinate system is given by the square-root of the corresponding metric coefficient for that axis. For obtaining the trajectory of any other object falling in X-Y plane on a gravitating body of mass M, we just need to locate the initial starting position of this object on the XY-T template graph and draw a geodesic of the required slope. However, it must be understood that for plotting trajectories of objects moving in the gravitational field of a body of different mass M', the differential scaling factor or the corresponding metric coefficients must be adjusted accordingly. If we find that drawing a 3D XY-T differential scale graph is physically difficult, we can just compute the data points for the required trajectory from the given template and that will serve the purpose. The required graphic trajectories can then be obtained with the aid of appropriate computer application programs.

We can extend this methodology for obtaining trajectories of objects moving in 3D physical space, in the gravitational field of a gravitating body of mass M. For this we can first obtain a differential scale 4D manifold XYZ-T as a template such that the Newtonian trajectories in the given gravitational field appear as geodesic curves in this template manifold. Of course, the differential scale or the metric of this template manifold will have to be correlated with the mass M of the gravitating body. Now, to obtain the trajectory of any other object in the given gravitational field, we can mark the initial starting position of the object in the template manifold and then compute the trajectory as a geodesic through that position. Isn't it wonderful to use the 4D XYZ-T manifold, with a suitable differential scale, as a template for obtaining trajectories of objects as geodesic curves? However, we will have to adjust the differential scale or the metric coefficients of this template manifold according to the mass M of the gravitating body. And this is precisely what is being done through EFE in the GR model. Further, to ensure a constant speed of light propagation in all coordinates, we can choose the 4D XYZ-T manifold as a Minkowski manifold with a differential metric!

What needs to be highlighted here is that in GR, the Riemannian 4D space-time manifold is being used precisely as a differential scale template for getting the trajectories of objects as geodesic curves. There is no doubt, what so ever, that GR is just a mathematical model used for obtaining the trajectories of objects as geodesic curves! However, the founders of GR did attempt to elevate this mathematical model to the status of a physical theory by assuming the 4D space-time manifold to be a physical spacetime continuum and also assuming that the mass-energy content of a gravitating body somehow controls the metric of this physical entity. Once we realize that 4D spacetime manifold is just an abstract mathematical construct and not a physical entity, then it is quite a simple matter to understand that GR is just a mathematical model and not a physical theory.

9

Invalidity of General Theory of Relativity

9.1 Spacetime Continuum & Postulate of General Relativity

Introduction. In the General Theory of Relativity (GR), a mathematical notion of space-time continuum is implied to be a physical entity that can get physically deformed or curved under the influence of gravitational field. The gravitational phenomenon is then modeled on the Riemannian geometry of this spacetime continuum. As per the current mathematical model of GR, the dynamic trajectories of material particles in a gravitational field are replaced by the geodesics in the deformed or curved spacetime continuum. The fundamental assumption of treating the spacetime continuum as a physical entity, is generally accepted on axiomatic footing and never analyzed on logical grounds.

Consider a very simple example of a particle motion in a circular orbit in an XY plane. The motion of this particle can be represented as a helical trace in an XY-T coordinate space or manifold. The velocity and acceleration characteristics of this particle will be represented by the geometry of helical trace in the XY-T manifold. An important point to be noted here is that the helical trace does not physically exist anywhere at any time; it is just a mathematical or graphical representation of the motion of a particle over a period of time. Similarly, the motion of various particles in three-dimensional physical space can be represented through suitable traces in four-dimensional XYZ-T space-time manifold. An important point to be noted here too is that four-dimensional traces of particles do not physically exist anywhere at any time; these are just mathematical representations of the motion of particles in three dimensional space over a period of time. In the same way, a four-dimensional space-time manifold XYZ-T does not physically exist anywhere at any time; it is just an abstract mathematical construct.

Postulate of General Relativity. The main postulate of General Relativity is that the gravitational phenomenon can be satisfactorily represented by 'suitably adjusting' the metric properties of the space-time manifold. For this the metric coefficients of the spacetime manifold are required to satisfy a set of partial differential equations (Einstein's Field Equations) involving energy-momentum tensor, whereby the non-rectilinear trajectories of mass particles will transform into geodesics. Thus the study of dynamical trajectories of mass particles in a gravitational field will reduce to the study of geodesics in the spacetime manifold defined by the specified metric. After doing so, the main postulate could be extended to imply that the gravitational field itself *somehow* modifies the metric coefficients of space-time manifold such that the trajectories of mass particles *naturally* turn out to be geodesics.

9.2 Line element 'ds' and the Space Metric

The most fundamental concept underlying all the basic notions of space is that of the absolute invariance of space points. Let us consider a particular space point P with coordinates (x^i) in a coordinate system X with origin at point O. If Q(x^i+dx^i) is another point in the neighborhood of P, then an infinitesimal separation distance ds between the points P and Q is given by

$$(ds)^2 = g_{ij}\, dx^i\, dx^j \qquad (9.1)$$

where g_{ij} are the metric coefficients in coordinate system x^i. The invariance of space points P, Q etc. implies that ds will remain constant, even when the coordinate system is transformed from X to say Y. Such transformations that ensure the invariance of the line element ds and hence the invariance of space points in general, are said to be admissible transformations. Obviously, if the metric coefficients $g_{ij}(x)$ constitute the metric of Euclidean space, then all other metric coefficients $h_{ij}(y)$ obtained through any admissible transformation of coordinates will also represent the metric of same Euclidean space.

On the other hand, let us consider an arbitrary change in metric coefficients $g_{ij}(x)$ in equation (9.1) to $h_{ij}(x)$. Here, by an arbitrary change we mean a change that is brought about on any considerations other than through an admissible transformation of coordinates. From equation (9.1), it can be easily seen that an arbitrary change in metric coefficients $g_{ij}(x)$ to say $h_{ij}(x)$, without any corresponding admissible transformation of coordinate parameters x^i, will lead to a change in separation distance ds between points P and Q to say ds'. Therefore,

$$(ds')^2 = h_{ij}\, dx^i\, dx^j \qquad (9.2)$$

Obviously, whenever the separation distance between neighboring points P and Q changes from ds to ds', it implies a relative shift in the original positions of P and Q to the changed positions say P' and Q' such that arc element P'Q' = ds'. This relative shift in positions of P and Q to the changed positions P' and Q' may be referred as the relative displacement of these points. Specifically, the vector PP' may be defined as the displacement vector **U** and the corresponding displacement of Q to Q' will then be represented by the incremented displacement vector **U+δU** (Fig. 9.1). The arbitrarily changed coefficients $h_{ij}(x)$ can be associated with the metric of a Riemannian space, whereas the original coefficients $g_{ij}(x)$ represented the metric of Euclidean space. Hence, whenever the metric $g_{ij}(x)$ of Euclidean space is changed to the metric $h_{ij}(x)$ of Riemannian space, the separation distance ds between two neighboring points P and Q will change to ds' as given by equations (9.1) and (9.2), with the associated displacement vector field **U** defined at all points P of the space continuum.

Fig. 9.1 Change in arc element ds to ds' produces displacement vector field **U**.

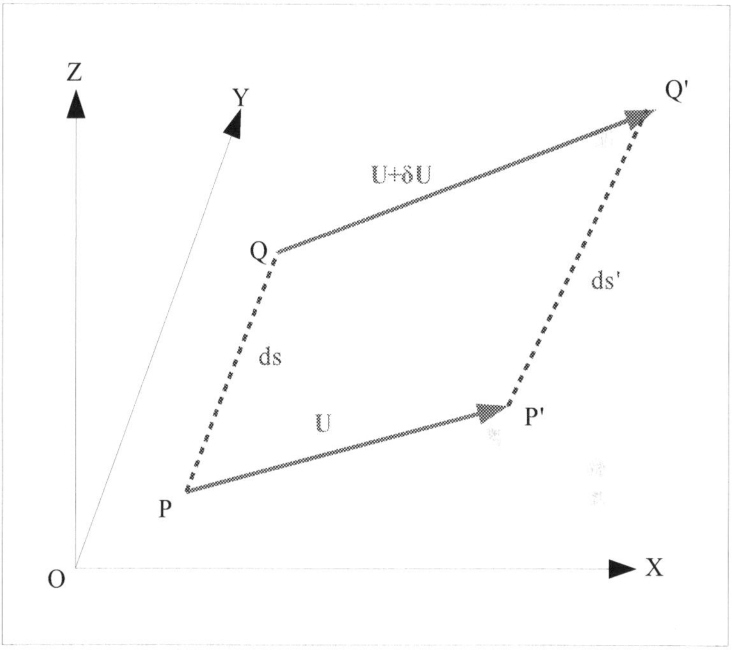

9.3 From Euclidean Space to Riemannian Space

Obviously the Euclidean and Riemannian geometries cannot be transformed into one another through admissible coordinate transformations without producing deformations in the associated space continuum. When a surface is represented in the parametric form by 2-D surface coordinates, the intrinsic geometry of the surface is described by its 2-D metric tensor. The Riemann tensor composed from the 2-D metric components is non-zero for a curved surface and zero for a plane surface. Let us critically examine the process under which a plane surface with Euclidean geometry can be changed over to a curved surface with Riemannian geometry.

Consider a large circular metal ring of radius R, filled inside with a plane thin film membrane (rubber membrane or soap film). The intrinsic geometry of any small region of this thin film can be represented by a 2-D flat metric with zero Riemann tensor. Let us now imagine that we exert a steady pressure over a small localized region of this film (say by impinging an air jet) in such a way that a small hemispherical bubble of radius $r \ll R$ is formed in this local region. The 2-D surface of this hemispherical bubble can be represented by a modified 2-D metric with non-zero Riemann tensor. Obviously, it is not difficult to visualize that the localized hemispherical bubble induced by a steady external pressure is actually a deformed (elongated/stretched) membrane with a curved surface in comparison to the undeformed plane membrane in the surrounding region.

By moving the impinging air jet sideways, the location of the hemispherical bubble on the large plane membrane can be easily shifted. The state of deformation of the curved membrane in comparison to the plane membrane can be studied in detail by comparing the Riemannian metric of the curved surface with the Euclidean metric of the plane surface. It can be easily shown that all displacements produced on the curved surface of the membrane are continuous and finite. The essential point to be stressed here is that a plane membrane surface with Euclidean metric ***does get deformed*** into a curved surface with Riemannian metric under the influence of external pressure. Precisely in the same way it has been postulated in GR that 'flat' space with Euclidean metric gets deformed to a curved space with Riemannian metric under the influence of a steady state gravitational field.

9.4 Deformed State of the Space Continuum

As per the fundamental postulate of GR, gravitational field of a gravitating body changes the metric $g_{ij}(x)$ of the surrounding Euclidean space to the metric $h_{ij}(x)$ of Riemannian space in accordance with Einstein's Field Equations (EFE). With this change in the metric of space continuum, the separation distance ds between any pair of neighboring space points P and Q will also change to ds' as given by equations (9.1) and (9.2). That means all space points within the region of gravitational field will experience relative displacements with the displacement vector field **U** producing the deformed or strained state of the space in that region. The strained state is fully represented by the strain tensor E with its components e^i_j given by the covariant derivatives of the displacement vector **U** as,

$$e^i_j = (u^i_{;j} + u^j_{;i})/2 \qquad (9.3)$$

In fact, subtracting (9.1) from (9.2) we get an important relation between the covariant components of the strain tensor e_{ij} and the metric coefficients as,

$$(ds')^2 - (ds)^2 = \{h_{ij} - g_{ij}\}\, dx^i\, dx^j = 2.e_{ij}\, dx^i\, dx^j \qquad (9.4)$$

$$\text{with} \quad 2.e_{ij} = h_{ij} - g_{ij} \qquad (9.5)$$

where e_{ij} represent the covariant components of strain tensor E expressed as functions of coordinate parameters x^i.

9.5 Gravitation Induced Metric & Associated Strain Tensor

Let us consider a spherical polar coordinate system with origin at point O and the coordinate parameters r, θ and ϕ. The metric coefficients for this coordinate system in the un-deformed or gravitation free space continuum are given as,

$$g_{rr} = 1 \ ; \quad g_{\theta\theta} = r^2 \ ; \quad g_{\phi\phi} = r^2.\mathrm{Sin}^2(\theta) \qquad (9.6)$$

The arc element or the separation distance ds between two neighboring space points P and Q in this region will be given by

$$(ds)^2 = g_{rr}(dr)^2 + g_{\theta\theta}(d\theta)^2 + g_{\phi\phi}(d\phi)^2$$
$$= 1.(dr)^2 + r^2.(d\theta)^2 + r^2.\mathrm{Sin}^2(\theta).(d\phi)^2 \qquad (9.7)$$

Now, let us assume that a spherically symmetric body of mass M and radius r_0, is located at the origin O of this coordinate system. Due to the gravitational field in its vicinity (i.e. $r > r_0 > 0$), the modified metric coefficients h_{ij} are given by the Schwarzschild solution as:

$$h_{rr} = 1/(1 - 2GM/c^2r) \ ; \quad h_{\theta\theta} = r^2 \ ; \quad h_{\phi\phi} = r^2.\mathrm{Sin}^2(\theta) \qquad (9.8)$$

Thus, the modified radial metric coefficient h_{rr} at any particular space point P(r, θ, ϕ) can be taken as a function of M and its value in the region under consideration is always greater than unity for M>0. The arc element or the modified separation distance ds' between two neighboring space point positions P' and Q' in this region will be given by

$$(ds')^2 = h_{rr}(dr)^2 + h_{\theta\theta}(d\theta)^2 + h_{\phi\phi}(d\phi)^2$$
$$= (1/(1 - 2GM/c^2r)).(dr)^2 + r^2.(d\theta)^2 + r^2.\mathrm{Sin}^2(\theta).(d\phi)^2 \qquad (9.9)$$

Therefore, using equation (9.5) we can compute the induced strain tensor components e_{ij} from the modified metric coefficients h_{ij} as,

$$2\, e_{rr} = h_{rr} - g_{rr} = (1/(1 - 2GM/c^2r)) - 1 \qquad (9.10)$$

with the factor $2GM/c^2r \ll 1$, equation (9.10) will get simplified to,

$$e_{rr} = GM/c^2r \qquad (9.11)$$

and $e_{\theta\theta} = h_{\theta\theta} - g_{\theta\theta} = 0 \;;\; e_{\phi\phi} = h_{\phi\phi} - g_{\phi\phi} = 0$ (9.12)

and $e_{\theta\phi} = e_{\phi\theta} = e_{r\phi} = e_{\phi r} = e_{r\theta} = e_{\theta r} = 0$ (9.13)

This set of strain tensor components constitutes the strain field induced in the region of space continuum where the gravitational field of M has modified the metric coefficients to h_{ij}. Further, it is quite interesting to note that from equation (9.9), the radial separation distance between two concentric spherical surfaces defined by radial coordinates $r = R_n$ and $r = R_n + dr$ is given by,

$$ds' = (1/(1 - 2GM/c^2R_n)^{1/2}).dr$$
$$= (1 + GM/c^2 R_n).\ dr$$

Here, if we take dr as one unit then

$$ds' = (1 + GM/c^2 R_n).$$

This shows that the radial separation distance ds' between two concentric spherical surfaces at radius R_n which are initially separated by a unit distance in the Euclidean space, will increase to $(1+ GM/c^2 R_n)$ in the Schwarzschild space.

9.6 Incompatibility of the Induced Strain Components

For a complete description of the strained state of the space continuum, we must be able to uniquely determine the displacement vector field **U** from the specified strain tensor components. For this the strain tensor components (e_{ij}) are required to satisfy Saint Venant's integrability or compatibility conditions. The displacement vector components obtained from the integration of partial differential equations of the type (9.3), must be single valued, finite and continuous functions of coordinates and must satisfy physical constraints over the boundary of the region of space under consideration.

It can be easily seen that the radial strain components e_{rr} given by equation (9.11), with all other components being zero, cannot satisfy the required compatibility conditions. In order to illustrate and highlight this problem, let us consider the relative displacement vector **U** that gives rise to the strain components e_{rr}, $e_{\theta\theta}$ and $e_{\phi\phi}$. If u^r is the only non-zero component of the displacement vector **U**, then the strain components dependent on u^r are given by,

$$e_{rr} = \partial u^r/\partial r \;;\; e_{\theta\theta} = u^r/r \text{ and } e_{\phi\phi} = u^r/r \quad (9.14)$$

Obviously, if the radial strain component e_{rr} is non-zero, the radial displacement component u^r must be non-zero. But once the radial displacement component u^r is non-zero, the tangential strain components $e_{\theta\theta}$ and $e_{\phi\phi}$ cannot be zero. This precisely is the incompatibility of the strain components e_{rr}, $e_{\theta\theta}$ and $e_{\phi\phi}$ induced by the static gravitational field of a spherically symmetric body of mass M. This incompatibility is not limited to the strain components induced by the Schwarzschild metric of spherically symmetric, static gravitational fields but is applicable to all strain components induced by the Riemannian metric obtained from EFE. In fact one of the essential requirements imposed by the standard compatibility conditions on strain components e_{ij} is that the Riemann tensor composed from e_{ij} must be a zero tensor. This can be true only if both metrics of equation (9.5), namely g_{ij} and h_{ij} are Euclidean which however contradicts the basic postulate of General Relativity. Therefore, the specification of metric coefficients (9.8) as per the Schwarzschild solution is physically invalid and unacceptable.

Further, even if we overlook the compatibility conditions for a while, from equations (9.11) and (9.14) we get,

$$e_{rr} = \partial u^r/\partial r = du^r/dr = GM/c^2r \tag{9.15}$$

which appears to be easily integrable. A simple integration of equation (9.15) yields the radial displacement component,

$$u^r = (GM/c^2).Ln(r/r_0) \tag{9.16}$$

Apparently, we seem to have obtained an elegant solution for the radial displacement component, in spite of the fact that the compatibility conditions were not satisfied by the strain components. Well, a closer look at equation (9.16) will show that this finite value of u^r will give rise to a finite value of tangential strain components $e_{\theta\theta}$ and $e_{\phi\phi}$ (equation (9.14)) whereas these components are required to be zero as per the modified metric h_{ij} (equation (9.12)) of the space continuum. This accounts for the incompatibility. Further the radial displacement u^r given by equation (9.16) tends to infinity as r tends to infinity. This is invalid since as per physical constraints, u^r must tend to zero when the gravitational field tends to zero at infinitely large 'r'.

9.7 Physical Invalidity of General Theory of Relativity

In an attempt to represent the gravitational phenomenon through geodesic equations in four dimensional space-time manifold, the main focus has been on the geometry of particle traces in the space-time continuum. For that, the space-time continuum had to be deformed in a particular way through EFE induced changes in its metric. However, instead of critically examining the induced deformations of the space continuum, the issue got buried under a misleading nomenclature of *space curvature*.

In GR, the structure of physical space continuum has been imprudently tampered with, by hypothesizing that the metric coefficients of space are affected by the presence of gravitational field. However, there is no physical mechanism available through which a gravitational field could force such a change in the metric of space. The EFE require the metric of space, under gravitational influence, to be inherently Riemannian. However, as seen above, when the metric of space is changed from Euclidean to Riemannian, the induced deformation of the space continuum gives rise to a mutually incompatible set of strain components leading to discontinuities and voids in the locations of neighboring space points. Hence, the induced Riemannian metric transforms the space continuum to space dis-continuum that is physically invalid and unacceptable.

Therefore, the main postulate of General Theory of Relativity is found to be logically invalid, mainly on account of,

- The mathematical notion of 4D spacetime continuum has been assumed to be a physical entity.
- The absence of any physical mechanism through which a gravitational field could force a change in the metric of space leading to its physical deformation.
- Due to the *incompatible deformation of space* induced by the Riemannian metric associated with gravitational field, the space continuum is implied to deform in to a dis-continuum of points.

Thus, irrespective of any claimed utility, application or validation of this theory, the General Theory of Relativity is logically ill founded, physically invalid and hence null and void. However, GR may still be treated as a mathematical model wherein an abstract spacetime manifold with differential scaling is used as a graphical template for obtaining the trajectories of objects as geodesic curves.

10

Uncertainty & Potential Energy in QM

10.1 Wave Particle Duality

According to the Heisenberg Uncertainty Principle in Quantum Mechanics (QM), there is a limit to the precision of measurement of the position and momentum of a particle at the same time. As per Wikipedia - "the uncertainty principle states that when measuring conjugate quantities, the product of their standard deviations must be at least $\hbar/2$". Here, conjugate quantities mean a pair of variables defined in such a way that they become Fourier transform duals of one-another. Fundamentally, the uncertainty principle is based on the wave-particle duality.

Evidence for the description of light as waves, was well established by the end of the 19th century and the photoelectric effect introduced firm evidence of a particle nature as well. On the other hand, the particle nature of electrons was well documented when the De Broglie hypothesis and the subsequent experiments by Davisson and Germer established the wave nature of the electron.

Wave-particle duality is deeply embedded into the foundations of QM. In the formalism of the theory, all information about a particle is encoded in its wave function - a complex function roughly analogous to the height of a wave at each point in space. This function evolves according to the Schrödinger equation, and this equation gives rise to wave-like phenomena such as interference and diffraction. The particle-like behavior is most evident due to the phenomena associated with measurement in QM. The measurement will return a well-defined position, a property traditionally associated with particles.

As per De Broglie's hypothesis, all microscopic particles of matter display a wave like nature while in motion. The De Broglie relations show that the wavelength λ is inversely proportional to the momentum p of a particle (λ=h/p) and that the frequency f is directly proportional to the particle's kinetic energy ($f=E_k/h$). This implies that the dynamic characteristics of a micro particle in motion, can be ascribed to the wave characteristics of the wavelet accompanying the particle. The intensity of the wavelet is expected to be large in the immediate vicinity of the particle and small elsewhere. It implies that the kinetic energy E_k of the moving particle is contained in the wave associated with the particle. Hence, to make the particle move, we have to supply a certain amount of energy E_k which gets stored in the wave associated with the particle.

10.2 Bohmian Mechanics

Quoting from the Stanford Encyclopedia of Philosophy - "Bohmian mechanics, which is also called the de Broglie-Bohm theory, characterized by the pilot-wave model and the causal interpretation of quantum mechanics, is a version of quantum theory discovered by Louis de Broglie in 1927 and rediscovered by David Bohm in 1952. It is the simplest example of what is often called a hidden variables interpretation of quantum mechanics. In Bohmian mechanics a system of particles is described in part by its wave function, evolving, as usual, according to Schrödinger's equation. However, the wave function provides only a partial description of the system. This description is completed by the specification of the actual positions of the particles. The latter evolve according to the 'guiding equation', which expresses the velocities of the particles in terms of the wave function. Thus, in Bohmian mechanics the configuration of a system of particles evolves via a deterministic motion choreographed by the wave function. In particular, when a particle is sent into a two-slit apparatus, the slit through which it passes and where it arrives on the photographic plate are completely determined by its initial position and wave function."[7]

10.3 Statistical Interpretation of the Wave Function

However, as per the Copenhagen interpretation of QM, the intensity of the wavelet is interpreted as the position probability density for the location of the micro particle. That is, in QM, the location of the center of a micro particle in motion is assumed to be smeared across the whole region of the wavelet as position probability density. The wavelet accompanying the micro particle in motion is described with a wave function $\psi(r,t)$ such that the position probability density $P(r,t)$ is given by,

$$P(r,t) = |\psi(r,t)|^2 \tag{10.1}$$

However, by interpreting the intensity of the wavelet as the position probability density of the micro particle and normalizing the wave function accordingly, the QM has effectively been transformed into Statistical Quantum Mechanics.

The assumption of interpreting the intensity of the wavelet as the probability density, is the point of departure of QM from the physical reality concerning the description of individual micro particles. Although in the Bohmian version of QM, the specification of the actual positions of particles is given by the guiding equation, still it shares with the mainstream QM the assumption of interpreting the intensity of the wavelet as the probability density. This interpretation for the intensity of the wave function, renders the QM description of the micro phenomenon as a statistical description. Therefore, it is quite logical to seek a statistical interpretation for the notion of uncertainty in QM.

10.4 Statistical Notion of Uncertainty in QM

The existence of the probability density P(r,t) makes it possible to calculate the expectation value of the position vector of the micro particle in motion. The expectation value is the mathematical expectation (or mean) in the sense of the probability theory. The expectation values of the position vector **r** and the momentum vector **p** can be written as,

$$\langle r\rangle = \int r\, P(r,t)\, dv \quad = \int \Psi' \, r \; \Psi \, dv$$

and

$$\langle p\rangle = \int p\, P(r,t)\, dv \quad = \int \Psi' \, p \; \Psi \, dv$$

The uncertainty in the expectation value of x component of position vector **r** is denoted by Δx and is defined to be the root-mean-square deviation from the mean value of x. Similarly, the uncertainty in the expectation value of **p** is denoted by Δp and is defined to be the root-mean-square deviation from the mean value of **p**. [8]

$$(\Delta x)^2 = \langle (x-\langle x\rangle)^2\rangle = \langle x^2\rangle - \langle x\rangle^2$$

$$(\Delta p)^2 = \langle (p-\langle p\rangle)^2\rangle = \langle p^2\rangle - \langle p\rangle^2$$

The product of these uncertainties can be shown to yield the famous uncertainty relation: $\Delta x . \Delta p \geq \hbar/2$.

This notion of uncertainty as the standard deviation from the mean values of salient parameters of the particle motion is statistical in character.

Uncertainty Assumptions. It may therefore be concluded that the famous uncertainty principle is founded on the following assumptions, some of which may not remain valid under all circumstances.

(a) As per de Broglie's hypothesis, all micro particles in motion are accompanied by a pilot wavelet, whose wave parameters are governed by the kinetic energy and momentum of the micro particle.

(b) The intensity of the wavelet is interpreted as the probability density for various parameters associated with the micro particle in motion.

(c) The uncertainty Δx or Δp in the expectation values of x or p is defined to be the root-mean-square deviation from the mean or expectation values of x or p.

Let us now examine the specific effect of the above assumptions. If the micro particle is not in motion, its momentum and kinetic energy will be zero and obviously therefore, there will be no pilot wavelet accompanying this particle. Hence assumption (a) is no longer valid for a micro particle at rest. Logically therefore, the uncertainty relations will not be applicable for any micro particle at rest. With assumption (b) the motion of individual micro particles is no longer governed by the de Broglie's wavelet containing kinetic energy and momentum of the particle, but by the normalized probability wave function. This renders the motion of individual micro particles as probabilistic and subject to provisions of the uncertainty principle. However, assumption (c) renders the uncertainty principle suitable for statistical applications only.

Uncertainty in the motion of individual micro Particles. In view of the specific assumptions associated with the formulation of the uncertainty principle, it can be shown that the uncertainty principle is no longer valid for the motion of individual micro particles.

- As per Bohmian mechanics actual positions of the micro particles are specified through the guiding equation resulting in a deterministic motion choreographed by the wave function. Hence, through the combined use of Schrödinger's wave equation and the guiding equation, a sequence of positions of the micro particle can be precisely determined that will constitute the particle trajectory **r**(t). Therefore the principle of uncertainty can no longer be applied to such a motion of individual micro particles.
- As per Heisenberg's Uncertainty Principle we cannot measure the position and momentum of a particle, with unlimited precision, at the same time. But we don't need to measure both the position and momentum of a particle at the same time. Let us measure only the position of a micro particle with dead precision at various instants of time t, thereby determining the particle trajectory r(t) to the desired precision. All other parameters of the particle motion (like momentum) can be subsequently computed from r(t) with any desired level of precision, without invoking the provisions of the uncertainty principle at any stage.

Hence, we may conclude that the uncertainty principle is essentially suitable for statistical applications and is not applicable for describing the individual motion of a micro particle.

10.5 Schroedinger's Equation for Free Particle

The Schroedinger's wave equation of Quantum Mechanics may be considered as founded in the suggestion of L. de Broglie that some sort of waves accompanied electrons and other micro particles in motion. These waves were assumed to represent the crucial dynamic characteristics of motion of the particle, namely the momentum **p** and total energy E, through following two relations adapted from the photon wave packet.

$$p = h/\lambda \tag{10.2}$$

$$E = h\,\nu \tag{10.3}$$

Here, h is the Planck's constant, λ the wavelength and ν the frequency of the above mentioned motion induced waves accompanying the particle. Of course, unlike the photon wave packet which as a whole behaves like a particle, the motion induced wave packet accompanying a material particle is a separate entity - an appendage to the particle. The material particle (i.e. a particle with non-zero rest mass) itself does not 'become' or transform into a wave packet. Let us consider the motion of a free particle, say an electron, located at point Q in a Cartesian coordinate system with center at point O as shown in figure 10.1. At any instant t let $\mathbf{r}_1$ be the position vector of point $Q(x_1,y_1,z_1)$ and let $|r_1|$ be the magnitude of this position vector. Since the particle at $Q(r_1)$ is assumed to be in motion, it will be accompanied by a motion induced wave packet spread over or extending into a region of space of say, volume τ, around the location of the particle.

Figure 10.1

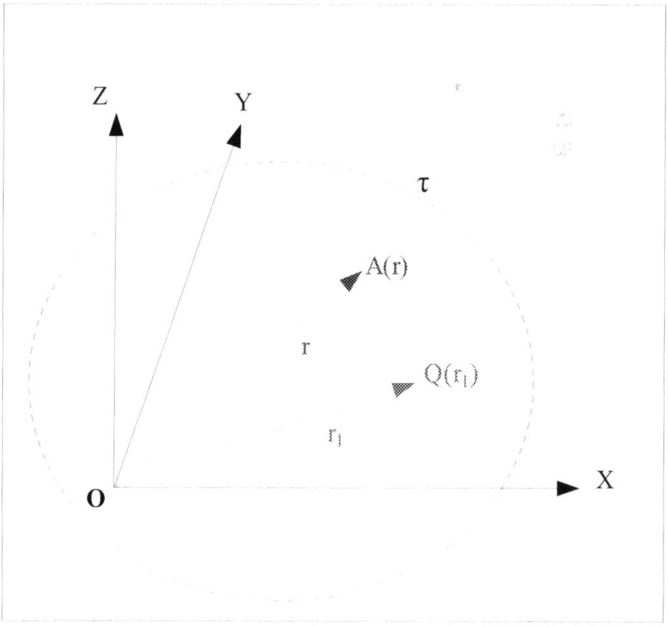

Let this motion induced wave packet be characterized by the parameter ψ known as wave function. The ψ is non-zero and finite at all points within the region 'τ' and vanishes at the boundary of τ. The whole region of space τ where the wave function ψ is defined or exists, may be termed as the ψ wave field. At any space point A(x,y,z), with position vector **r**, the value of wave function ψ will keep varying with time t and hence may be written as $\psi(\mathbf{r},t)$. The $\psi(\mathbf{r},t)$ in general will be a complex function of space and time coordinates. The intensity of the wave function will be given by

$$|\psi|^2 = |\psi.\psi^*| = P(\mathbf{r},t) \tag{10.4}$$

The wave intensity $P(\mathbf{r},t)$ is also known as the probability density. The wave function ψ is normalized by equating the integral of $P(\mathbf{r},t)$, over the entire volume τ of the ψ wave field, to unity. The intensity or probability density $P(\mathbf{r},t)$ is expected to be maximum in a region of space where the moving particle is actually expected to be located at that instant of time. If we consider a small element of volume $\delta\tau$ around point $A(\mathbf{r})$ where the probability density is $P(\mathbf{r},t)$, then the actual probability of locating the particle $Q(\mathbf{r}_1)$, within this volume $\delta\tau$ is given by $[P(\mathbf{r},t).\ \delta\tau\ ]$.

At any particular instant t, the point $A(\mathbf{r})$ may be considered as a general field point within the wave field τ and the point $Q(r_1)$ may be considered as a fixed point representing the instantaneous location of the particle within this ψ wave field. During the motion of this particle, the position vector $\mathbf{r}_1$ of point Q will keep changing with time. That is, $\mathbf{r}_1$ can be expressed as a function of time t.

$$\mathbf{r}_1 = \mathbf{f}(t) \tag{10.5}$$

Therefore, equation (10.5) will represent the trajectory of the moving particle. At the given instant, let us examine the wave function $\psi(\mathbf{r},t)$ within the entire wave field τ when the moving particle is located at point $Q(r_1)$. This wave function $\psi(\mathbf{r},t)$ is obtained as a solution of Schroedinger's wave equation. The Schroedinger's equation in turn is derived from the energy conservation principle as applied to the moving particle, by making use of the following two operators which form the core of Quantum Mechanics.

$$\mathbf{p} \rightarrow -\boldsymbol{i}\ \hbar\ \nabla \tag{10.6}$$

$$E \rightarrow \boldsymbol{i}\ \hbar\ \partial./\partial t \tag{10.7}$$

where $\boldsymbol{i} = \sqrt{(-1)}$ and $\hbar = h/2\pi$. Using the total energy (E) relation for a free particle of mass (m), with kinetic energy (T) and potential energy (V)

$$E = T + V \tag{10.8}$$

and with V= 0 ;

$$E = p^2/2m + 0 \qquad (10.9)$$

Multiplying equation (10.9) with ψ and applying the operators (10.6) and (10.7), we get,

$$i\,\hbar\,\partial\psi/\partial t = -(\hbar^2/2m).\nabla^2\psi \qquad (10.10)$$

This is the well known Schroedinger's wave equation for a free particle.

10.6 Notions of Kinetic, Potential and Total Energy

Kinetic Energy. If the kinetic energy (T) of a free particle is reduced to zero by bringing it to rest, we can see from equations (10.6) to (10.10) that $\nabla\psi$, $\nabla^2\psi$, $\partial\psi/\partial t$ and ψ will all reduce to zero. That is, in the absence of kinetic energy, the ψ wave field of the particle will collapse to zero. It implies that the ψ field of a moving particle is essentially dependent on the kinetic energy of the particle. Hence, the kinetic energy must be existing as field energy of the associated wave field or contained within the ψ field of the particle. In fact, the intensity of the normalized wave function $P(\mathbf{r},t)$ or $|\psi|^2$ given by equation (10.4), may actually represent normalized kinetic energy density of the particle. This also corresponds to the concept of energy density in electromagnetic field being proportional to the squares of electric and magnetic field strengths.

We can generally say that any change in the motion of a particle will induce a corresponding change in the kinetic energy of that particle and vice-versa. Extending this notion to the ψ wave field of a moving particle, we can say that any change in the motion of the particle will induce a corresponding change in the overall ψ wave field of that particle and vice-versa. That is, any change in the overall ψ wave field of a particle will induce a corresponding change in the motion of that particle. Thus, we may appreciate that high energy interactions of micro particles could be governed by the superposition interactions of their ψ wave fields. This precisely may be the reason of phenomenal success of QM in the study of micro particle interactions.

Potential Energy. Let us consider the case of an isolated electron. The electrostatic field of an electron, with radially decaying electric field strength, can be identified with its wave field with radially decaying amplitude. A significant portion of the mass energy of the electron is actually stored or contained in this electrostatic or wave field. The wave field energy component of the electron mass is an integral part of the electron and is not dependent on the existence of any other charge or field in its vicinity. Now, let us consider a proton and electron pair separated by distance d from each other. Their respective positive and negative electrostatic fields or wave fields will get overlapped or superposed almost throughout their spatial extension. Consequently, the combined field energy of the proton-electron system, being proportional to the square of the resultant field strength, will be slightly less than the total sum of the individual field energies of the isolated charges. This reduction in the combined field energy of the proton-electron system, is precisely the negative interaction energy due to the Coulomb interaction of the system and is known as the negative potential energy of the proton electron pair. Therefore,

Potential energy of proton-electron pair = $V = -e^2/4\pi\varepsilon_0 d$ (10.11)

or,

Interaction energy released by the system = $|V| = e^2/4\pi\varepsilon_0 d$ (10.12)

Obviously, the interaction energy released by the proton-electron pair is zero when they are separated by an infinite distance. As their separation is reduced to a distance d, the interaction energy released by the system, as given by equation (10.12), is continuously converted into the kinetic energy of the proton-electron pair. Of course, during this conversion or transfer of released interaction energy to the kinetic energy of the system, the overall conservation of energy, momentum and angular momentum is automatically ensured. As already noted, the kinetic energy of a particle is stored or contained in the ψ wave field of the particle. Therefore, the potential energy or the interaction energy of the proton-electron pair may be considered as a parameter signifying the transfer of energy between the ψ wave field and the combined field energy of the system. Thus, a +ve potential energy of a particle signifies the transfer of ψ wave field energy (i.e. K.E.) to the field energy (i.e. mass energy) of the particle. Similarly, a -ve potential energy signifies the transfer of a part of the field energy to the ψ wave field energy of the interacting particles. The Schroedinger's wave equation is intended to describe the variations in ψ wave field of a moving particle as a result of such energy transfers.

It is therefore obvious that the potential energy of an electron with respect to a proton at distance d, represented as V(d), can not be regarded as a field parameter in the sense that it does not represent any entity distributed in space. For example, the electrostatic field strength or wave field energy density can be regarded as field parameters because they represent the entities which are existing or defined at all space points of the associated field at any instant of time. On the other hand, potential energy is the interaction energy depending entirely on relative location of the electron with respect to the proton at any particular instant and is not defined or existing at any other space point at that instant.

However, the term potential energy, is not applicable for a single isolated particle due to absence of any interaction. It has a meaning only for two or more interacting particles or fields, wherein transfer of energy could take place between the combined electrostatic or wave field energy and the total ψ wave field energies of the interacting particles or fields, without violating the principle of conservation of momentum and energy. If at any instant t, the proton (considered almost stationary) is located at point O, the origin of coordinate system (Fig. 10.1) and the moving electron is located at point $Q(r_1)$ with position vector $\mathbf{r}_1$, then the potential energy of the electron will depend on the magnitude of r_1 and represented by $V(r_1)$. It will not be a function of the coordinates of field point A(r). That is, when the wave function is represented as $\psi(r,t)$, the potential energy term can not be represented as V(r).

Total Energy. The total energy E of a system of two interacting particles (the proton-electron pair in the present case) is defined through equation (10.8) , that is,

$$E = T + V(r_1) = T + [- e^2/4\pi\varepsilon_0 r_1] \qquad (10.13)$$

Truly speaking, the term E in equation (10.8) or (10.13) represents the total external energy supplied to the system. E is +ve when this amount of energy is externally added or supplied to the system of interacting particles and is -ve when it is extracted, or taken out of the system. The externally supplied energy may either get added to the kinetic energy of the system or to their electrostatic field energies through the potential energy term and vice-versa. Generally a negative E will represent a bound state of the system of interacting particles. Even though the name implies total energy, it actually does not include the mass energies of the interacting particles. Therefore, a constant total energy E or a stationary energy state of an isolated system of interacting particles, implies the constancy of sum of K.E. and potential energy of the system.

10.7 Schroedinger's Equation with Wrong P. E. Term

The total external energy, whether supplied to or extracted from the system will naturally influence the kinetic energy T and hence the ψ wave field of the moving particle. The complex relationship between the variations of total and potential energy, and the corresponding space-time variations of the ψ wave field representing the kinetic energy, is reflected through the Schroedinger's wave equation involving the potential energy term. Multiplying equation (10.13) with $\psi(\mathbf{r},t)$ throughout and then applying the operators (10.6) and (10.7), we get,

$$E \cdot \psi(\mathbf{r},t) = (p^2/2m) \cdot \psi(\mathbf{r},t) + V(r_1) \cdot \psi(\mathbf{r},t) \qquad (10.14)$$

and $$i\,\hbar\,\partial\psi/\partial t = -(\hbar^2/2m).\nabla^2\psi + V(r_1) \cdot \psi \qquad (10.15)$$

But the Schroedinger's wave equation is normally written in the form,

$$i\,\hbar\,\partial\psi/\partial t = -(\hbar^2/2m).\nabla^2\psi + V(r).\psi \qquad (10.16)$$

Equations (10.15) and (10.16) differ in the P.E. terms $V(r_1)$ and $V(r)$.

In the Schroedinger's original wave equation (10.16), the potential energy is expressed as a function of the coordinates of general field point $A(\mathbf{r})$, instead of the coordinates of instantaneous location $Q(\mathbf{r_1})$ of the particle (Figure 10.1). That means the Schroedinger's wave equation (10.16) is founded on the total energy relation

$$E = T + V(r) = T + [-e^2/4\pi\varepsilon_0 r] \qquad (10.17)$$

instead of equation (10.13). As already discussed above, the potential energy of an electron-proton pair is strictly a function of their instantaneous relative distance r_1 and is not defined at any other space point $A(\mathbf{r})$. This discrepancy is not a simple or inadvertent mistake in the Schroedinger's wave equation (10.16) but rather a serious conceptual mistake with far reaching consequences. This mistake is continued with throughout Quantum Mechanics, where the potential energy term $V(r)$ is often replaced by $e.\phi(r)$; with scalar potential $\phi(r)$ treated as a function of coordinates of general field point $A(\mathbf{r})$ rather than a function of coordinates of instantaneous location $Q(\mathbf{r_1})$ of the particle. The greatest temptation for permitting this mistake, might have been the consequent ease of solving the Schroedinger's equation (10.16) by treating the potential energy term $V(r)$ as spherically symmetric and independent of time. Even though most weaknesses of Quantum Mechanics could be attributed to this conceptual mistake, yet for want of timely rectification, the mistake had to be 'swept under the uncertainty carpet'.

10.8 Consequential Errors in Established Solutions

Let us now examine a few consequential errors in the established solutions of Schroedinger's wave equation (10.16), arising out of the above mentioned mistake in the potential energy term $V(r)$. For this, let us consider the ground state 1s orbital of Hydrogen atom, the normalized wave function $\psi_{nlm} = \psi_{100}(\mathbf{r},t)$ of which is,[9]

$$\psi_{100}(\mathbf{r},t) = (1/\pi a_0^3)^{1/2}.\exp(-r/a_0).\exp(-i\,E_1 t/\hbar) \qquad (10.18)$$

where, $E_1 = - 13.6$ ev is the total energy of the 1s orbital
and $a_0 = 0.53$ $\mathring{A}$ is the corresponding Bohr radius.

Oscillating ψ Wave Field. In accordance with the original suggestion of L. de Broglie, the solution for ground state 1s orbital of Hydrogen was expected to yield some sort of waves or wave packet, accompanying the orbiting electron. But equation (10.18) represents a spherically symmetric standing wave oscillations of the ψ field, which does not correspond to the physical situation. Equation (10.18) can not represent any traveling wave or a wave group which could describe the orbiting motion of the electron. Hence, fundamentally this solution is unsuitable to represent the physical situation and should have been rejected. This error could be attributed to the wrong potential energy term V(r) used in the Schroedinger's equation (10.16), as discussed above.

Spherically Symmetric ψ Wave Field. As discussed earlier, at any instant the intensity of ψ wave function is expected to be maximum in the vicinity of location of the electron at that instant. But from equation (10.18) we get,

$$P(\mathbf{r},t) = |\psi_{100}|^2 = (1/\pi a_0^3).\exp(-2r/a_0) \qquad (10.19)$$

Equation (10.19) shows that the intensity of the ψ wave function is spherically symmetric with its maximum value at the center where the proton is located. The obvious conclusion from this result could be that the electron and proton are both located at the center, which of course is physically impossible. The physical situation demanded that the ψ wave packet should not only have accompanied the orbiting electron but also should have been centered at and spread around the instantaneous location of the electron.

Therefore, it could be concluded that solution (10.18) does not represent the physical situation and should have been rejected as invalid. This error too could be attributed to the wrong potential energy term V(r) used in the Schroedinger's equation (10.16). It is, of course, a different matter that use of correct potential energy function $V(r_1)$ might have rendered the Schroedinger's equation (10.15) analytically unsolvable. Ideally speaking, the Schroedinger's equation (10.15) should be solved in conjunction with pre-set initial and boundary conditions to yield a complete solution consisting of:

(a) The electron trajectory $\mathbf{r_1} = \mathbf{f}(t)$

(b) The wave function $\psi(\mathbf{r}, \mathbf{r_1}, t)$ or $\psi(\mathbf{r}, t)$.

In fact the electron trajectory can be computed even without Schroedinger's wave equation. Therefore, we must reject equation (10.18) as an invalid solution.

Zero Orbital Angular Momentum of the Electron. The ground state 1s Hydrogen orbital solution $\psi_{100}(\mathbf{r},t)$ given by equation (10.18), corresponds to zero orbital angular momentum. The zero angular momentum, by virtue of its basic definition and its fundamental physical concept, will represent either a stationary electron or an electron passing through the nucleus. Both of these alternatives do not correspond to the physical situation of an orbiting electron and hence invalid. This error also seems to have occurred due to the wrong potential energy term V(r) used in the Schroedinger's equation (10.16).

Negative Kinetic Energy of the Electron. As per the usual terminology in QM, equation (10.18) represents the ψ wave function for the lowest stationary state of electron in Hydrogen atom. In this stationary state, the total energy eigenvalue E_1 is equal to -13.6 ev. For this stationary state, the probability density P(r) is given by equation (10.19). The integral of this probability density over the entire ψ field (i.e. for r varying from zero to infinity) works out to unity, as expected, since ψ is normalized. This result is interpreted as the overall probability of finding the electron within the entire ψ field is 100 % . Let us now work out the overall probability of finding the electron within a spherical shell of inner radius r_1= 0.5 a_0 and outer radius r_2 = 1.5 a_0 where a_0 is the Bohr radius.

$$P_{12}=\int_0^{2\pi}\int_0^{\pi}\int_{r_1}^{r_2} P(r).r^2\,\mathrm{Sin}(\theta)\,dr\,d\theta\,d\phi$$

$$=4\pi\int_{r_1}^{r_2} P(r).r^2 dr$$

= 0.4965 ..... by using equation (10.18)

That means the overall probability of finding the electron within a spherical shell of radii 0.5 a_0 and 1.5 a_0 is 49.65 %. We may interpret this result as follows. If somehow we could conduct a very large number of electron location determining experiments, then in about 49.65 percent cases we are likely to locate the electron within a spherical shell of radii 0.5 a_0 and 1.5 a_0 . Apparently this is quite a reasonable result and in the backdrop of uncertainty principle it is generally accepted as true.

Now let us carry out one more computation. This time let us work out the overall probability of finding the electron outside a sphere of radius $r_3 = 2a_0$. Proceeding on the same lines as above, we get this probability as,

$$\mathbf{P}_3 = 4\pi \int_{r_3}^{\infty} P(r).r^2 dr \qquad = \frac{4}{a_0^3} \int_{2a_0}^{\infty} e^{(-2r/a_0)}.r^2 dr \quad = \quad 0.2381$$

That means the overall probability of finding the electron outside a sphere of radius $2a_0$ is about 23.81 percent. A closer look will show that this result is totally wrong. For this, let us once again consider the parent equations from which this result is derived, namely the Schroedinger's equation (10.16) and the total energy equation (10.17). Since the result (10.18) is valid in the region outside a sphere of radius $2a_0$, the equations (10.16) and (10.17) must also be valid. However, for a stationary state represented by equation (10.18), the total energy

$E_1 = -13.6$ ev $= -2.176\times10^{-18}$ J,

is known to be constant. Therefore, from equation (10.17) we get,

$$T = E_1 - [- e^2/4\pi\varepsilon_0 r] = e^2/4\pi\varepsilon_0 r - 2.176\times10^{-18} \text{ Joules} \qquad (10.20)$$

This shows that kinetic energy T of the electron, in the stationary state (10.18), keeps reducing with increasing r. Equation (10.20) shows that T reduces to zero at $r = 2a_0$. That means when the electron is located outside a sphere of radius $2a_0$, its kinetic energy will become negative. But we have seen above that probability of finding the electron outside a sphere of radius $2a_0$ is 23.81 percent. Hence, we draw the conclusion that as per original Schroedinger's equation (10.16), there is 23.81 % probability that the electron, in ground state of Hydrogen atom, will exist in a negative kinetic energy state with imaginary velocity components. Since this is patently an absurd conclusion, we must review the situation. Therefore, we come back to our previous observation that the potential energy term V in the original Schroedinger's equation (10.16) has been wrongly taken as a function of coordinates of general field point A(**r**), instead of taking it as a function of coordinates of point $Q(r_1)$, the instantaneous location of the electron.

From equation (10.20), we can easily see that for the ground state of Hydrogen, the electron will always remain bound well within a bounding sphere of radius $2a_0$. That is, whatever be the shape of trajectory or electron orbits ($\mathbf{r}_1 = \mathbf{f}(t)$) in the ground state of Hydrogen, these orbits will always be located well within this bounding sphere of radius $2a_0$, i.e. $r_1 < 2a_0$. Now, let us assume for the time being that we succeed in obtaining analytical solutions of correct Schroedinger's equation (10.15). Then the ψ wave field will have to be restricted within this bounding sphere, so that the overall probability of finding the electron outside this sphere is zero. If however, the ψ wave field could not be reduced to zero outside this bounding sphere, the wave function intensity $|\psi|^2$ will have to be re-interpreted as the normalized kinetic energy density instead of electron probability density.

Concluding Remarks. In spite of the error in potential energy term V, as brought out above, the QM has been extremely useful in the study of high energy micro particle interactions. Of course, the limitations of QM are generally not highlighted. As for example, the atomic and molecular orbitals are assumed to describe only the time averaged charge density distributions around nucleus and not the trajectories of electrons. However, for those applications where either the effect of potential energy term is negligible or the potential energy function V is made more dependent on particle locations, the QM has been of immense value. It is hoped that through rectification of the error in potential energy term V in the Schroedinger's wave equation, we may be in a better position to further enhance the efficacy and utility of QM. Possibly, the already complex wave function ψ might ultimately turn out to be much more complex.

Therefore, the necessity of strengthening the logical foundations of Quantum Mechanics, so as to bring it within the grasp of imagination, cannot be overemphasized. After all, from a purely philosophical standpoint, the end result of any intellectual pursuit must come within the mental grasp, within the perceptible limits of the human mind. That is, any deeper understanding of the physical reality of the micro world which we may develop through an intellectual process by using simple or mathematical logic, must be fully perceptible to the human mind.

11

The Hydrogen Orbitals

11.1 Orbital motion of an Electron around a Proton

Introduction. To develop a clear mental picture of electron orbits in Hydrogen atom, we first need to develop the fundamental concepts of the electron and its electrostatic field. In order to visualize the instant to instant motion of an electron orbiting a proton, we must understand as to how exactly the two charges interact to release potential energy, where the K.E. is stored and how the photon is created. Hence, our basic approach in this chapter will be to first make use of some of the most fundamental concepts about electron, nature of charge, field energy and field interaction. With these fundamental concepts, we shall analyze the energy balance of an isolated proton-electron system and develop the electron trajectory by using the energy and angular momentum conservation principle. Since the subject analysis constitutes a theoretical study, and since no physical measurements on the instantaneous positions of the electron are envisaged in this study, the postulate or principle of uncertainty is not applicable here.

We shall derive a relation from purely classical considerations that by emitting a photon at angular frequency ω, the angular momentum of orbiting electron is changed by $\hbar$ due to mechanical recoil action. This fact will form the basis for quantization of angular momentum and hence total energy in Hydrogen orbitals. Further we shall retain the use of quantum numbers **n, ℓ, m** as usual. However, on the considerations of restricting the change in angular momentum to $\hbar$, we shall associate quantum number ℓ with angular momentum of $(\ell+\frac{1}{2}).\hbar$ instead of $\sqrt{(\ell(\ell+1))}.\hbar$. This will lead to all elliptical electron orbits. During the emission of a photon, the elliptical orbit transitions at constant angular frequency ω will also be computed and plotted.

As we shall see in chapter 18, the electron structure consists of a spherically symmetric standing strain wave core surrounded by a radial phase wave type electrostatic field. About 35% of the total mass energy of the electron (or positron) is distributed in its electrostatic wave field. The Coulomb interaction between two charge particles is effected through superposition of their wave fields. Positive interaction energy between two similar charges implies the transfer of a portion of their kinetic energies to their combined field energy. Negative interaction energy between two dissimilar charges implies the transfer of a portion of their combined field energy to their kinetic energies. Of course, the total energy and momentum of the system is conserved in both cases. Without going into the internal details of the electron structure, let us examine the effect of motion on the overall wave field of the electron.

Motion Induced Fields and Kinetic Energy. Let us consider uniform motion of an electron along +x axis, at velocity v. Due to the finite velocity c of the phase waves, the complete wave field of the electron will get deformed. This field deformation may be considered through the concept of retarded time and retarded position vector leading to motion induced change in amplitude and direction of phase waves at any particular point. Since the kinetic energy of the moving particle will be stored in its deformed field, most of the field components are expected to be increased. Let us say that the original field vector is deformed such that the change is defined as motion induced field vector **A**, which will vanish when the particle velocity becomes zero. The motion induced electric and magnetic fields of the moving particle can now be derived from the time derivative and curl of this induced field as:

$$\mathbf{E} = -(1/\varepsilon_0).(1/c).\partial\mathbf{A}/\partial t \qquad (11.1)$$

and

$$\mathbf{B} = (\mu_0.c).[\nabla\times\mathbf{A}]\,. \qquad (11.2)$$

Under certain conditions of motion, some part of the induced fields could be dissociated from the moving particle, whereas the bound field can never be dissociated unless the particle gets annihilated. The induced fields are an integral part of the moving particle system and it is a matter of interpretation whether the particle motion controls the induced fields or the induced fields govern the particle motion. We shall use these concepts for developing electron orbits in the Hydrogen atom.

Isolated Proton - Electron System. Let us now consider an isolated proton - electron system with the proton located at the center of chosen coordinate system. Neglecting the motion of proton as too small, we consider the constrained motion of the electron under conservation of system energy and angular momentum. Initially, when the electron is far removed from the proton, let its kinetic energy (T) and electrostatic potential energy (-V) be negligible or zero. We adopt a sign convention that all symbols like T, V, E etc. representing energy will always be positive. Conventionally, the total system energy -E, with electron far removed, is considered zero (-E = -V+T = 0). But, in reality, we know that the total energy of the system does include the mass energies of the two particles. When the electron is brought to a finite radial distance r from the proton, without any external work, the conventional -ve potential energy (-V) of the system gradually changes from zero to $-e^2/4\pi\varepsilon_0 r$ or say $-\eta/r$ where $\eta=e^2/4\pi\varepsilon_0$ is a constant. Simultaneously, the kinetic energy T of the electron increases from zero to η/r. That is, the interaction energy released by the system, keeps getting converted to kinetic energy of the electron on an instant to instant basis. Henceforth, we shall no longer use the term negative potential energy (-V) but only refer to field interaction energy released (V) by the system. If a small part (E_n) of this energy is now radiated out as a photon, then the system total energy will become $-E_n$ and the remaining K.E. of the electron is given by $T=V - E_n$.

Figure 11.1 Interaction energy released (V) by the Electron - Proton system vs. their relative radial distance R. In the elliptical orbital motion of the electron, the relative radial distance oscillates between OC and OB with corresponding K.E. varying between C_1C' and B_1B'.

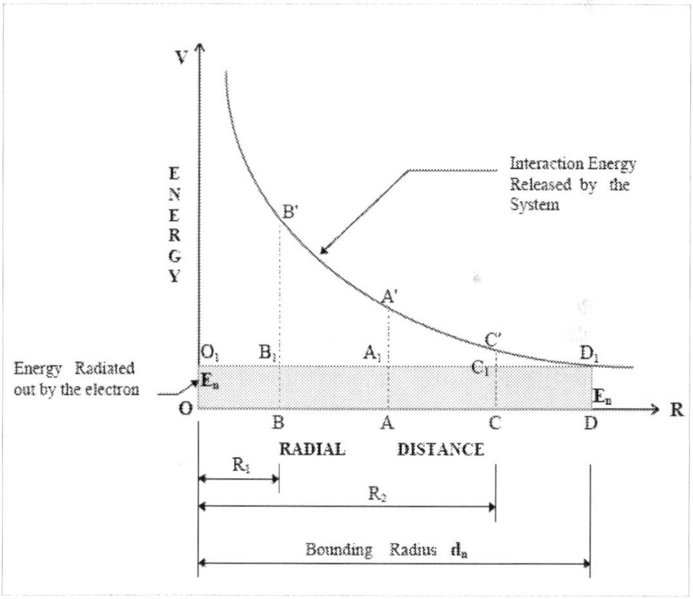

Time Invariant Orbital Parameters. Corresponding to the conventional total energy of $-E_n$, let d_n be the radial distance at which the K.E. of the electron (mass m_e) becomes zero (i.e. $V= E_n$), as shown in figure 11.1. A sphere of radius d_n may be referred as the bounding sphere for the electron. All possible electron orbits for angular momenta $k.\hbar$, must be located well within this sphere characterized by principle quantum number **n**. Let us therefore examine the shape and size of all possible orbits for given orbital quantum numbers **n** and ℓ. There is one unique circular orbit of radius $a_c = d_n/2$, angular momentum $L_c = \mathbf{n}.\hbar$, with following other main parameters,

$$T_c = V - E_n = \eta.(1/a_c - 1/d_n) = \eta/d_n = E_n \quad (11.3)$$

$$L_c^2 = (m_e.v_c.a_c)^2 = 2m_e.a_c^2.T_c = 2m_e.a_c^2.E_n \quad (11.4)$$

$$v_c = 2T_c/m_e v_c = (\eta/\hbar\mathbf{n}) \text{ and } v_c = L_c/m_e.a_c = (\mathbf{n}.\hbar/a_c m_e) \quad (11.5)$$

which give,

$$a_c = \mathbf{n}^2.(\hbar)^2/m_e\eta \text{ and } \mathbf{E_n} = \eta/2a_c = \mathbf{(m_e/2n^2).(\eta/\hbar)^2} \quad (11.6)$$

However, with the same total energy, there could be many elliptical orbits with their angular momentum $L_e = k.\hbar < L_c$ and differing from each other in steps of $\hbar$. Since $L_e=0$ will correspond to head on collision and annihilation of electron, it cannot correspond to any valid orbit. Therefore, as mentioned earlier, we shall take $k=(\ell+½)$ instead of $(\ell(\ell+1))^{½}$. Let the two vertices B and C (fig. 11.2) of the ellipse be identified by subscripts 1 and 2, e.g. radius R_1, R_2 etc. Then, at the vertices:

$$L_e = k.\hbar = m_e.v_i.R_i \quad (\text{for } i = 1, 2 \text{ and no summation on i})$$

$$T_i = ½.m_e.v_i^2 = \eta.(1/R_i - 1/d_n) = E_n.(d_n/R_i - 1) \quad \text{which gives}$$

$$R_i = E_n.d_n/(T_i+E_n) = 2a_c.E_n/(T_i+E_n) \quad (11.7)$$

and $L_e^2 = 2m_e.T_i.R_i^2 = 8m_e.a_c^2.E_n^2.T_i/(T_i+E_n)^2 = 4L_c^2.E_n.T_i/(T_i+E_n)^2$

which, after substituting $L_e/L_c = k/\mathbf{n}$ simplifies to a quadratic in T_i as,

$$T_i^2 - 2[2(\mathbf{n}/k)^2-1].E_n.T_i + E_n^2 = 0 \quad (11.8)$$

that yields two values of T_i, that is T_1 and T_2 as given below

$$T_1 = E_n.[(2(\mathbf{n}/k)^2-1) + 2(\mathbf{n}/k).((\mathbf{n}/k)^2-1)^{½}] \quad (11.9)$$

$$T_2 = E_n.[(2(\mathbf{n}/k)^2-1) - 2(\mathbf{n}/k).((\mathbf{n}/k)^2-1)^{½}] \quad (11.10)$$

and from equation (11.7), R_1 & R_2 can now be computed. Further, from these values of R_1, R_2, T_1, T_2, we can compute maximum and minimum values of velocity (v_1, v_2) and angular frequency $\omega_1 = v_1/R_1$, $\omega_2 = v_2/R_2$. It can also be easily shown that ellipse major diameter = $R_1+R_2 = d_n$ and that the eccentricity of the ellipse is given by $e_k = (R_2 - R_1)/d_n$.

11.2 Dynamic Orbital Parameters

After determining the major diameter $2a_c$ the eccentricity e_k & R_2, v_2, ω_2 etc. for the given quantum numbers **n**, ℓ, we are now in a position to compute the instant to instant motion of the electron on this orbit. For this purpose, let us introduce Cartesian coordinate system X-Y with origin Q at the center of major diameter BC as shown in fig. 11.2.

Figure 11.2 In the orbital motion of the electron as its relative radial distance from the Proton O varies from OC to OB the angle β increases from 0 to π. Major dia. d_n of the ellipse is governed by **n** and the eccentricity by ℓ.

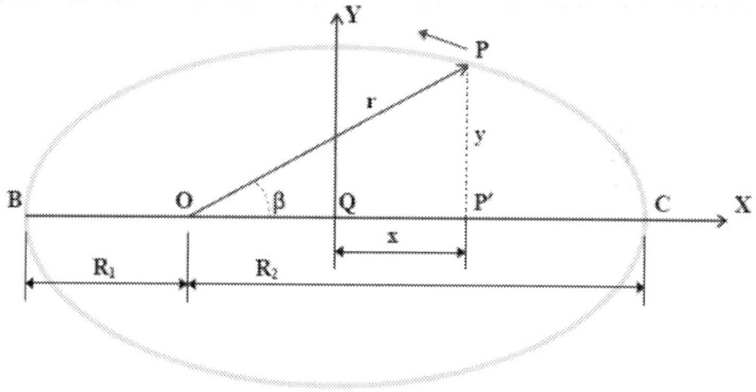

Of course, the radial position vector **r** will be measured from O, the principal focus of the ellipse (proton location). At time t =0 let us start from the outer vertex C, where $r = R_2$, $x = a_c$, $y = 0$, $\beta = 0$, $v_\beta = v_2$, $v_r=0$ and $\omega = \omega_2$. For numerical computations with the aid of a digital computer, we may divide the major diameter BC into N (say 1000) equal parts such that $\delta x=d_n/N$. To compute the next position (subscript n) parameters from the old position, the following relations could be used:

$$x_n = x_o - \delta x \qquad r_n = a_c + e_k.x_n$$

$$y_n = (1 - e_k^2)^{1/2}.(a_c^2 - x_n^2)^{1/2} \qquad \beta_n = \mathrm{Sin}^{-1}(y_n / r_n)$$

$$v_\beta = (\kappa.\hbar)/m_e r_n \qquad \omega_n = v_\beta / r_n = (\kappa.\hbar)/m_e r_n^2$$

$$\delta\beta = \beta_n - \beta_o \qquad \delta t = \delta\beta/\omega_n$$

$$t_n = t_o + \delta t \qquad v_r = e_k.\ \delta x/\delta t$$

$$v = (v_r^2 + v_\beta^2)^{1/2} \qquad (11.11)$$

Repeating these computational steps, we can obtain instant to instant variation of all dynamic parameters like v, β, ω, r, etc. and plot them against time. For convenience in computing, it is preferable to use non-dimensional form of these parameters by dividing with corresponding parameters of the first circular Bohr orbit. A few typical curves showing v and ω verses time for certain elliptical orbits are shown in figures 11.3 & 11.4. The results indicate the necessity of reviewing our concepts about orbital stationary states in QM. By assuming separable temporal part $e^{(-i.E_n.t/\hbar)}$ in the wave function, basically circular electron orbits get implied in the concept of stationary states, whereas in reality the quantum numbers **n** and *ℓ* yield all elliptical orbits.

Figure 11.3 Velocity - time graph for 4p electron.

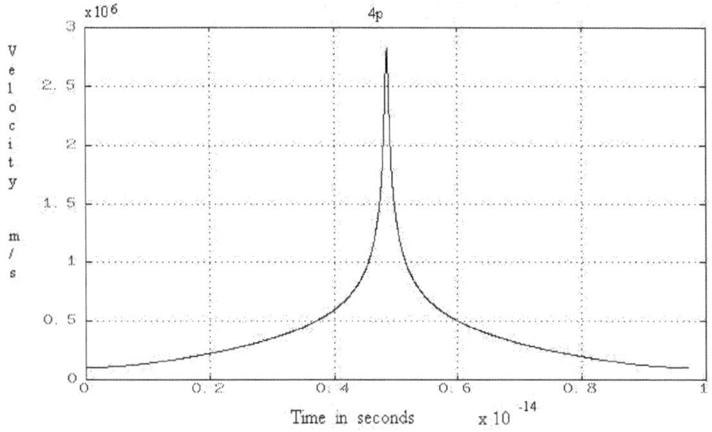

Figure 11.4 Angular frequency – time graph for 4p electron.

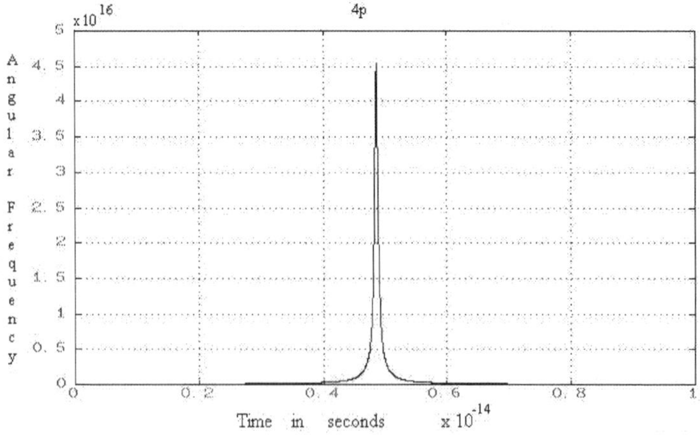

The salient orbital parameters for Hydrogen are given at Table-1. One very important parameter to be noted from this table is the time period of the orbital motion, which is independent of ℓ. That is, for a given **n** the time period of all elliptical orbits is the same as that of a corresponding circular orbit. This may explain why equivalent circular orbits could be implied in the concept of stationary states.

Table - 1. Salient Orbital Parameters

Orbit No.	Total energy	Eccen-tricity	Time Period	Vertex- Radii		K.E. at Vertices		Angular Frequency	
				Min.	Max.	Max.	Min.	Max.	Min.
	ev		10^{-15} s	10^{-10} m		ev		10^{15} rad/s	
1s	- 13.6	0.866	0.1520	0.071	0.988	189.44	0.977	1151.15	5.93
2s	- 3.40	0.968	1.2164	0.067	4.167	210.77	0.055	1280.73	0.33
2p	- 3.40	0.661	1.2164	0.717	3.518	16.69	0.693	33.80	1.40
3s	- 1.51	0.986	4.1052	0.067	9.461	214.59	0.011	1303.95	0.06
3p	- 1.51	0.866	4.1052	0.638	8.889	21.05	0.109	42.64	0.22
3d	- 1.51	0.553	4.1052	2.130	7.397	5.25	0.435	6.38	0.53
4s	- 0.85	0.992	9.7308	0.066	16.871	215.92	0.003	1312.03	0.02
4p	- 0.85	0.927	9.7308	0.618	16.319	22.45	0.032	45.47	0.07
4d	- 0.85	0.781	9.7308	1.858	15.080	6.90	0.105	8.39	0.13
4f	- 0.85	0.484	9.7308	4.369	12.569	2.45	0.296	2.12	0.26

11.3 Photon Emission - General Conditions

In accordance with the foregoing discussions, we may visualize the emission of a photon wave packet from the vicinity of an orbiting electron, under the following general conditions:

(a) The angular frequency ω=kc, will govern the spatial extension as well as the energy content of the photon wave packet.

(b) The angular frequency ω should remain constant throughout the spatial and temporal extension of the wave packet.

(c) A photon of angular frequency ω_p may be emitted from the induced field of orbiting electron when the instantaneous angular frequency

ω of the orbital motion matches ω_p and remains constant throughout the emission process.

(d) The photon may be emitted from a finite region around the middle of the relative position vector **r** of the orbiting electron.

(e) The photon will be emitted in the orbital plane of the electron and along a line perpendicular to the relative position vector **r**.

(f) The strength and time rate of change of induced **E** and **B** fields must be above certain minimum values, to enable the emission of a photon wave packet of certain minimum energy content ($\hbar\omega_p$).

(g) The strength of induced **E** and **B** fields in the region around **r**/2, will be governed by $\mathbf{v}_r$ and $\mathbf{v}_\beta$ components of the electron velocity respectively. The time rate of change of these fields may be governed by instantaneous angular frequency ω of the orbiting electron.

(h) The direction of emission of the photon may get reversed if the relative phase of induced **E** field is opposite at the time of emission. That is, the photon will be emitted in the direction of $\mathbf{v}_\beta$ if the electron is approaching the nucleus ($\mathbf{v}_r$ -ve) and in a direction opposite to $\mathbf{v}_\beta$ if the electron is receding ($\mathbf{v}_r$ +ve) from the nucleus at the time of emission.

(i) During the photon emission process, the conservation of overall system energy and momentum must be ensured at every instant.

Photon Emission - Recoil. Let us consider a body A with kinetic energy E and another body B with negligible K.E., both located on and moving along X-axis. Let the body A act on body B for a small distance δs to transfer a small fraction δE of its K.E. to body B. We may assume that the action force **F** exerted by body A and the reaction force -**F** exerted by B, remains constant throughout this energy transfer interaction between A and B. Then the energy transferred from A to B will be given by $\delta E = F.\delta s$. If this interaction process lasts for a very small interval of time δt, then a momentum impulse of $I_p = F.\delta t$ will be imparted to both A and B in opposite directions along X-axis. This impulse will imply a small change in momentum δp such that,

$$\delta p = I_p = F.\delta t = (\delta E/\delta s).\delta t = \delta E / (\delta s/\delta t) \qquad (11.12)$$

Now, let us imagine that the body A mentioned above is the orbiting electron and body B is the photon being emitted. The recoil impulse experienced by the electron while transferring a small fraction δE of its K.E. to the photon is therefore given by equation (11.12). Since, as mentioned above, the photon is emitted in a direction perpendicular to the position vector **r**, the distance traveled by the electron δs may be given by r.δβ so that δs/δt=r.ω. The corresponding change in angular momentum of the orbiting electron is therefore given by,

$$|\delta L| = r\times\delta p = r.\delta E/r\omega = \delta E / \omega \ . \qquad (11.13)$$

Hence, the total change in angular momentum ΔL, when a photon of total energy content $\Delta E = \omega.\hbar$ is emitted, will be,

$$|\Delta L| = \Delta E / \omega = \omega.\hbar/ \omega = \hbar \qquad (11.14)$$

This is an important result which forms the basis of angular momentum quantization and hence total energy quantization in sub-atomic phenomena. The photon emission recoil phenomenon is unique in two respects. Firstly, the actual recoil interaction between the electron and the photon is effected through the action of released photon fields $\mathbf{E}_p$ and $\mathbf{B}_p$ on the bound -ve electrostatic field of the moving electron. At the same time the released photon fields $\mathbf{E}_p$ and $\mathbf{B}_p$ will act on the bound +ve electrostatic field of the nucleus to produce an opposite linear momentum change. The presence of the nucleus will thus ensure that a total momentum of $\Delta p = \Delta E/r\omega$ is not carried by the photon but only a small fraction $\Delta E/c$ is carried off.

Secondly, the emission phenomenon is unique in the sense that depending on the relative phases of $\mathbf{E}_p$ and $\mathbf{B}_p$ fields, the photon may be emitted by the moving electron either in forward direction or in rearward direction. The photon will be emitted in forward direction when at the time of emission, the electron is approaching the nucleus on its elliptical orbit and the total angular momentum of the electron will reduce by $\hbar$. The photon will be emitted in rearward direction when the electron is receding from the nucleus and the total angular momentum of the electron will increase by $\hbar$.

11.4 Orbital Transition Parameters

Let us consider the electron transition from orbit A specified by $(\mathbf{n}_1, \ell_1)$ to orbit B $(\mathbf{n}_2, \ell_2)$, such that $\mathbf{n}_1 > \mathbf{n}_2$ and $\ell_2 = \ell_1 - 1$, then

$$E_1 = (m_e/2n_1^2).(\eta/\hbar)^2 , \qquad d_1 = \eta/E_1 ,$$

$$E_2 = (m_e/2n_2^2).(\eta/\hbar)^2 , \qquad d_2 = \eta/E_2$$

and the angular frequency of the photon to be emitted is $\omega_p = (E_2 - E_1)/\hbar$. Angular momentum for orbit A is $L_a=(\ell_1+½).\hbar$ and for orbit B is $L_b=(\ell_2+½).\hbar$. The emission will take place when the electron is approaching the nucleus. The complete orbital parameters for A and B can be worked out as per procedure outlined above. The radius R_a in orbit A, at which the photon emission process will commence, can be computed from the condition,

$$\omega = \omega_p = L_a/m_e R_a^2 \quad \text{or} \quad R_a = (L_a / m_e\omega_p)^{1/2}$$

$$\text{and similarly,} \quad R_b = (L_b / m_e\omega_p)^{1/2} \qquad (11.15)$$

While computing the dynamic parameters of orbit A, we may extract the values of salient parameters at the instant when $r = R_a$ and designate them with subscript o as β_o, T_o and t_o=0, r_o=R_a, L_o=L_a and $E_o = E_1$. Similarly, from orbit B we may extract β_b when $r = R_b$. To compute the transition trajectory from A to B we may divide this path into N (say 1000) equal steps. With $\delta L = \hbar/N$ and $\delta E = (E_2 - E_1)/N$, following relations may be used for step by step computation of transition trajectory.

$$L_n = L_o - \delta L \qquad E_n = E_o + \delta E$$
$$r_n = (L_n / m_e\omega_p) \qquad T_n = \eta/r_n - E_n$$
$$v_\beta = r_n.\omega_p \qquad v = (2T_n/m_e)^{1/2}$$
$$v_r = (v^2 - v_\beta^2)^{1/2} \qquad \delta r = r_n - r_o$$
$$\delta t = \delta r/v_r \qquad t_n = t_o + \delta t$$
$$\delta\beta = \omega_p .\delta t \qquad \beta_n = \beta_o + \delta\beta \qquad (11.16)$$

Figure 11.5 Electron Orbit Transition from 2p to 1s.

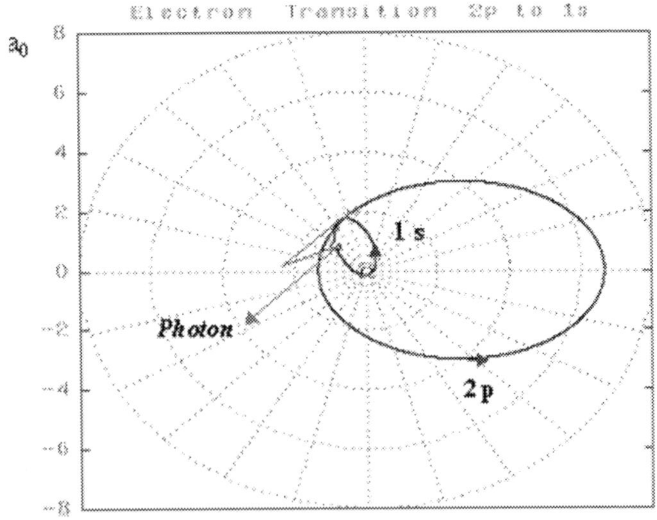

Figure 11.6 Electron Orbit Transition from 4d to 2p.

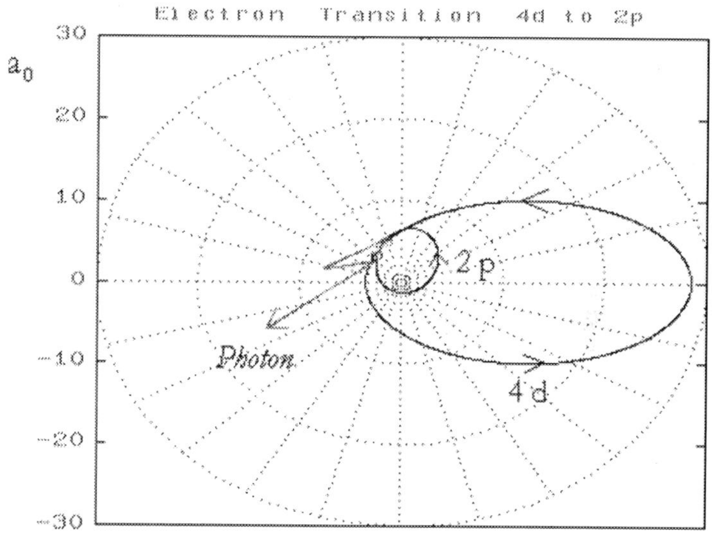

Figure 11.7 Electron Orbit Transition from 3s to 2p.

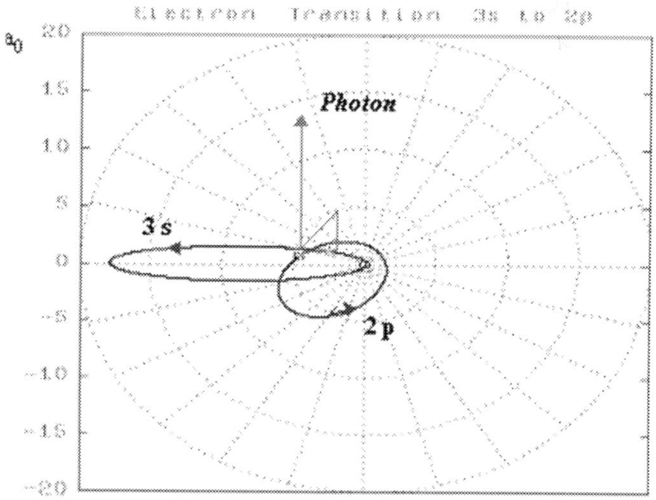

After repeating these steps N times, let the final values of t_n and β_n be t_f and β_f respectively. For plotting the transition trajectory along with orbits A and B, the major axis of orbit B will have to be rotated through $\Delta\beta = \beta_f - \beta_b$. A few plots of some typical orbital transitions are shown in figures 11.5 to 11.7 , showing original and new orbits in the same plane, along with the direction of photon emission. The photon emission time for various transitions is generally found to be of the order of 10^{-16} seconds which appears to be too small for the actual spatial extension of the photon. This is due to the fact that the photon is not 'created' from a single point in space, but 'released' from the spatially extended induced field of the electron and re-forms to its characteristic shape in accordance with the vector wave equation.

11.5 Summary & Conclusion

In this chapter we have attempted to develop a new model, a new methodology, to compute the detailed instant to instant motion of individual electrons in Hydrogen atom, based on the principle of conservation of energy and momentum. For this, a number of new basic concepts have been used to develop a better insight and fundamental understanding of the sub-atomic phenomenon. The new concepts include the structure of the electron, Coulomb interaction, potential energy etc. We have also shown that whenever a photon is emitted from an orbiting electron the angular momentum of that electron is changed by $\hbar$. This may be seen as the origin of various quantization rules. After introducing several new fundamental concepts, the electron trajectories in the form of elliptical orbits, have been developed and their transitions plotted. The linear velocities, angular velocities, K.E., radial distance **r** and angle β have been computed for the instant to instant motion of the electron in various Hydrogen orbitals.

12

M M X : A New Hypothesis on Photon Emission

12.1 M M Experimental Setup

The historic significance of Michelson Morley Experiment (MMX) is due to the belief that it had proved the non-existence of aether medium. The original MM experiment was conducted in the backdrop of 'aether wind' which was expected to influence the propagation of light, viewed as a continuous wave motion. Let us re-examine the MM experiment without any preconceived notions of 'aether wind'. Further, we need not view the propagation of light as a continuous wave motion. Instead, we may regard individual photons as small wave packets which propagate in empty space either independently or in unison with other similar wave packets. When a large group of similar wave packets propagate in empty space in unison or in synchronization, they may give the semblance of a continuous wave motion. However, the interaction and propagation characteristics of such a continuous wave will be governed by the characteristics of individual wave packets and we shall mainly focus on that. Let us review the same old MM experiment conducted with a stream of photon wave packets propagating in empty space or vacuum.

Let us consider the MMX setup as shown in figure 12. 1. Here, A_0 is the starting point from where a monochromatic light beam is split into two mutually perpendicular beams. One of these is directed along A_0X_0 direction and the other along A_0Y_0 direction. X_0 and Y_0 are two reflecting mirrors, such that:

$$|A_0X_0| = |A_0Y_0| = D\ .$$

After reflection from mirrors X_0 and Y_0, the two beams are recombined at A_0 to study the interference fringes. The whole setup is mounted on a rotating platform M. Let us position this MMX setup (platform M) in a region of empty space or matter free space or vacuum with following essential properties.

- The vacuum or empty space offers no drag or resistance to the motion of photon particles or wave packets through it.
- The vacuum is homogeneous and isotropic. That is, its properties are uniform all over and in all directions. That means the effect of vacuum on a moving particle or wavelet will be identical along all directions.

We shall critically examine the circumstances that lead to the null result of this experiment as obtained by Michelson & Morley and also attempt to provide a suitable explanation for this null result. Let us consider a Universal Reference Frame as discussed in chapter 5. Let OXY be a reference coordinate frame that is fixed relative to the above Universal Reference Frame.

Figure 12.1

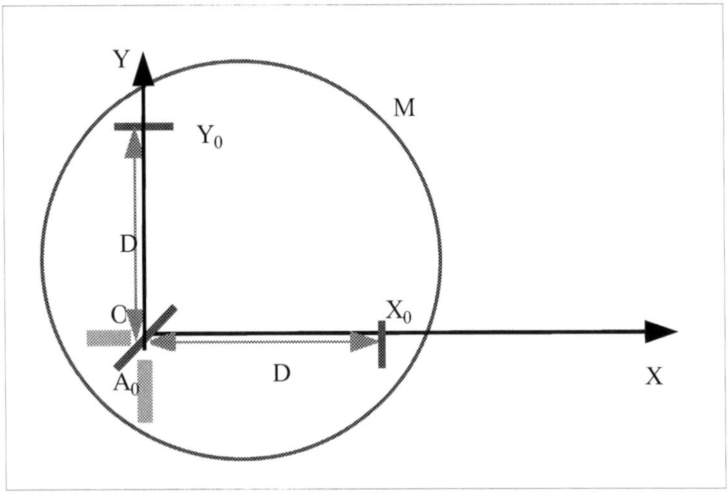

Case I : Experimental Platform M at Rest in OXY.

To begin with, at time t_0=0, let the point A_0 coincide with the origin O of the OXY reference coordinate frame. Let the arm A_0X_0 align with OX axis and arm A_0Y_0 align with OY axis as shown in fig. 12.1. When the platform M is at rest in OXY during the MM experiment, two monochromatic light beams consisting of photons of frequency f and wavelength L_0 will set off from A_0 at speed c towards X_0 and Y_0. After getting reflected from X_0 and Y_0 the two beams will arrive back at A_0 at the same instant of time (t=2D/c). The number of photon wave lengths covered on the path length of each beam (N_x and N_y) will be the same (f.t). That is, with reference to the wave packet just being transmitted, the change in phase angle of both reflected waves will be same.

$$\phi_x = 2\pi.N_x = 2\pi\, f.t = 4\pi D/L_0 \qquad (12.1)$$

$$\phi_y = 2\pi.N_y = 2\pi\, f.t = 4\pi D/L_0 \qquad (12.2)$$

Figure 12.2

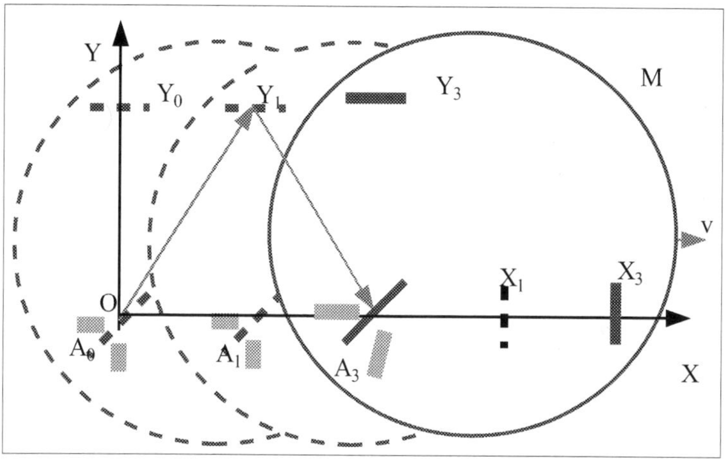

Thus, there will be no interference fringes (other than those due to inadvertent path differences in the actual experimental set up) and the rotation of the platform will not produce any fringe shift. The number of photon wave lengths N_x and N_y covered on the path length of each beam will also be directly proportional to the number of photons lined up on that path length. Further, when the two beams recombine, if the two wave packets happen to be in phase then the combining photons will get synchronized and the light beam will appear brighter. If the two wave packets are out of phase then the photons will scatter and the light beam will appear dimmer.

Case II : Platform M moves along OX at uniform velocity v.

In this case, at instant $t_0=0$, two light beams will set off from point A_0 at speed c in the directions X_0 and Y_0 respectively. Let us call the beam set off towards Y_0 as the transverse beam and the other one moving towards X_0 as the axial beam. After an interval of time t_1, the platform M will move a distance $v.t_1$ along OX such that the point A_0 moves to a new position A_1 on OX as shown in fig. 12.2. Simultaneously, the reflecting mirror locations X_0 and Y_0 would have moved to their new positions X_1 and Y_1 respectively such that:

$OA_1 = v.t_1$

and, $Ox_1 = D + v.t_1$

After an interval of time t_3 from the beginning, the platform M will move a distance $v.t_3$ along OX such that the point A_1 moves to a new position A_3 on OX as shown in fig. 12.2. Simultaneously, the reflecting mirror locations X_1 and Y_1 would have moved to their new positions X_3 and Y_3 respectively such that:

$OA_3 = v.t_3$ and, $Ox_3 = D + v.t_3$

Let us take the time intervals t_1 and t_3 to be such that the transverse beam of light, starting from point A_0 at time $t_0=0$, travels through vacuum or empty space at a uniform speed c, and gets reflected from the mirror at point Y_1 at time t_1. After reflection it arrives back to the receiver interferometer at point A_3 at time t_3 as shown. From the symmetry of the outward and inward light paths, we get $t_3=2t_1$. From triangle OA_1Y_1,

$$D^2 + (v.t_1)^2 = (c.t_1)^2$$

Or

$$t_1 = \frac{D}{\sqrt{(c^2-v^2)}} = \frac{D}{c} \cdot \frac{1}{\sqrt{1-v^2/c^2}} \tag{12.3}$$

and

$$t_3 = 2\,t_1 = \frac{2\,D}{c} \cdot \frac{1}{\sqrt{1-v^2/c^2}} \tag{12.4}$$

The number of wave lengths covered on the path length $A_0Y_1A_3$ of the transverse beam will be $N_y = f.t_3$. That is with reference to the wavelet at the beginning of this path length, the change in phase angle ϕ_y of the reflected wave at the end of this path will be,

$$\phi_y = 2\pi.N_y = 2\pi.f.t_3. \tag{12.5}$$

If L_0 is the wavelength (= c/f) of axial and transverse beam wave packets, then:

$$N_y = \frac{2.D}{L_0} \cdot \frac{1}{\sqrt{1-v^2/c^2}} \tag{12.6}$$

$$\phi_y = \frac{4\pi.D}{L_0} \cdot \frac{1}{\sqrt{1-v^2/c^2}} \tag{12.7}$$

Now let us assume that the axial beam of light starting from point A_0 at time $t_0=0$, also travels through vacuum or empty space at a uniform speed c and gets reflected from the axial mirror after a time interval t_2. At this instant t_2, whole platform M would have moved a distance $v.t_2$ along X axis and the points A_0, X_0 would have moved to positions A_2 and X_2 as shown in fig. 12.3. The time interval t_2 can be calculated from the relation,

$$OX_2 = OA_2 + A_2 X_2$$

That is, $$c.t_2 = v.t_2 + D$$

or $$t_2 = \frac{D}{(c-v)} \quad (12.8)$$

Let us further assume that the axial beam of light, after reflection from X_2 travels back at uniform velocity c and at instant t_4 reaches the interferometer, which by now has shifted to A_4 as shown in fig. 12.3.

$$X_2 A_4 = O X_2 - O A_4$$

Or $$c.(t_4 - t_2) = v.t_2 + D - v.t_4$$

Or $$t_4 - t_2 = \frac{D}{(c+v)}$$

Hence,

$$t_4 = t_2 + \frac{D}{c+v} = \frac{D}{c-v} + \frac{D}{c+v} = \frac{2D}{c} \cdot \frac{1}{(1-v^2/c^2)} \quad (12.9)$$

Figure 12.3

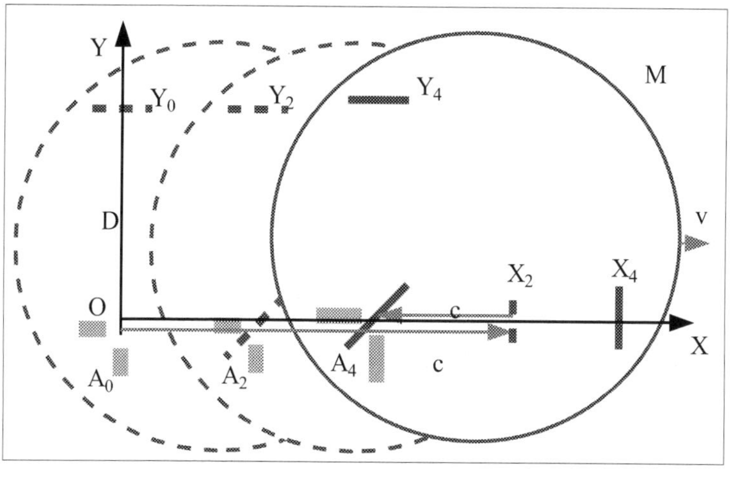

The number of wave lengths covered on the path length $A_0X_2A_4$ of the axial beam will be $N_x = f.t_4$. That is with reference to the wavelet at the beginning of this path length, the change in phase angle ϕ_x of the reflected axial wave at the end of this path will be,

$$\phi_x = 2\pi.N_x = 2\pi.f.t_4. \qquad (12.10)$$

With $L_0 = c/f$ for the axial beam wave packets,

$$N_x = \frac{2.D}{L_0}.\frac{1}{(1-v^2/c^2)} \qquad (12.11)$$

$$\phi_x = \frac{4\pi.D}{L_0}.\frac{1}{(1-v^2/c^2)} \qquad (12.12)$$

From equations (12.6) and (12.11), we can compare the number of wavelengths covered on the transverse and axial beam path lengths. These numbers N_y and N_x are also proportional to the number of photons lined up along the transverse and axial beam path lengths respectively. Clearly $N_x > N_y$ which indicates that more number of photons get accumulated on the axial path length $A_0X_2A_4$ than on the transverse path length $A_0Y_1A_3$. Therefore when the reflected transverse and axial beams recombine at the receiver interferometer they will no longer be in phase and produce interference fringes. The magnitude of their phase angle difference is given by equations (12.7) and (12.12). Therefore, from the values of ϕ_y and ϕ_x we observe that the transverse and axial beams of light, after reflection arrive at the receiver interferometer out of phase. The magnitude of difference between their phase angles on arrival is given by,

$$\phi_x - \phi_y = \frac{4\pi.D}{L_0}.\left(\frac{1}{(1-v^2/c^2)} - \frac{1}{\sqrt{(1-v^2/c^2)}}\right) \qquad (12.13)$$

which yields, for $v^2/c^2 << 1$,

$$\phi_x - \phi_y = \frac{4\pi.D}{L_0}.\left((1+\frac{v^2}{c^2}...)-(1+\frac{1}{2}\frac{v^2}{c^2})...\right)$$

$$= \frac{2\pi.D}{L_0}.(\frac{v^2}{c^2}) \qquad (12.14)$$

12.2 Null Result of M M Experiment

This magnitude of phase difference between transverse and axial light beams on arrival at the receiver will yield an observable fringe shift at the interferometer when the MM setup platform M is rotated by 90° or more. But when this experiment was actually conducted by Michelson & Morley, no significant fringe shift could be observed. This result implied that regardless of the relative velocity of platform M with respect to coordinate system OXY, the axial and transverse beams actually always reach back in phase. The absence of a significant fringe shift is regarded as a null result of the MM experiment. This null result had so far been explained through the main postulates of the Special Theory of Relativity. But with the invalidation of SR, we need to have a fresh look at the null result of the MM experiment and seek an alternative explanation for the same.

At the time the MM experiment was conducted, light was assumed to be a continuous wave motion through the aether. As we know now, light consists of a stream of discrete particles called quanta of light or photons. These photons possess finite amount of electromagnetic field energy and momentum. Just like the motion of other discrete particles, the motion of discrete photon particles too must be controlled by their energy and momentum. The interference and reflection characteristics of beams of photons will be governed by the characteristics of the individual photons. Specifically, we need to examine the implied assumption regarding independence of photon frequency from the state of motion of the emitter or reflector. Since the thermal broadening of spectral lines is a well known phenomenon, it is quite obvious that the state of motion of the emitter or reflector does actually influence the photon frequency differently in different segments of the two light paths. This fact has not been taken into account in the interpretation of the MMX, and hence it constitutes a major flaw in the resulting inferences.

12.3 Flaw in the Interpretation of MMX Null Result

A scientific hypothesis refers to a provisional idea whose merit requires evaluation. For proper scientific evaluation of hypothesis, an appropriate experiment needs to be carefully designed and conducted. Such an experiment is often designed to produce a measurable physical output in the event of a true or valid hypothesis. However, if such an experiment yields a null result, it is invariably interpreted as implying a false or invalid hypothesis. Logically, such an implication of a null result is often misplaced and wrong.

In the design of an appropriate experiment for evaluation of certain hypothesis, a number of assumptions are often required to be made. Some of these assumptions are explicit which are considered valid as a matter of fact. However, some assumptions are implicit which are either not considered or their validity taken for granted as axiomatic.

Positive outcome of the experiment is based on the strict condition that all of these assumptions must be valid in the physical environment under which the experiment is being conducted. Let us assume that four assumptions A_1, A_2, A_3 and A_4 are involved in an experiment designed to evaluate a particular hypothesis H_1. The outcome of the experiment may be expressed as a logical statement as,

If A_1 is True & A_2 is True & A_3 is True & A_4 is True & H_1 is True;
then the experiment yields a Positive result.

On the other hand,

If A_1 is False Or A_2 is False Or A_3 is False Or A_4 is False Or H_1 is False;
then the experiment yields a Null result.

Now we can easily appreciate that a Positive result of the experiment will definitely confirm the truth or validity of the hypothesis H_1 under evaluation. But it is more important to appreciate that a null result of the experiment can never definitely confirm the invalidity of the hypothesis under evaluation, simply because the null result could also be produced by any of the assumptions not being true. Hence, if an experiment designed to evaluate a particular hypothesis yields null result, we must discard the experiment as a failure but logically cannot confirm the invalidity of the hypothesis.

Let us now apply this logic to the null result of MMX. The hypothesis intended to be evaluated by MMX may be stated as,

H_1 : The speed of light is constant 'c' only in the 'aether fixed' absolute reference frame and will appear to be different for different observers in relative uniform motion. Accordingly, it should be possible to measure the velocity 'v' of an observer or his local reference frame w.r.t. the aether fixed reference frame by measuring the speed of light in the observer's local frame.

The explicit and implicit assumptions made in the design of MMX for testing the hypothesis H_1 can be stated as:

- A_1 : Speed of light in vacuum is not influenced by the state of motion of the emitter.
- A_2 : Monochromatic light beam is fundamentally treated as a continuous wave phenomenon. The detailed emission process and interaction characteristics of individual photons do not influence the phase shift in different light paths of the MMX.

- A_3 : Frequency and wavelength of light is not influenced by the state of motion of the emitter.
- A_4 : Frequency of light does not change after reflection from mirrors in motion. Hence, frequency and wavelength of light in different light paths is not influenced by the state of motion of the MMX setup.

The null result of MMX logically implies that: Either A_1 is False Or A_2 is False Or A_3 is False Or A_4 is False Or H_1 is False.

Hence a null result of MMX can only be treated as a failed experiment without any implied inference regarding the success or failure of the main hypothesis under test. The same conclusion applies to all other MMX type experiments, designed to test Hypothesis H_1, where the final outcome of the experiment is a null result. Therefore, we must consider the MMX and all other similar experiments, to be the flawed experiments, the erroneous interpretation of which mislead the physics community to the wrong track.

12.4 A New Hypothesis on Emission of Photons

Let us imagine a space ship moving at a uniform velocity **u** m/s in the universal or absolute reference frame X. If the space ship is to eject a module of mass m kg in the forward direction such that the module attains a uniform velocity **V** m/s in reference frame X, then it will have to be given a forward impulse of m.(**V**- **u**) Newton seconds. That is, the momentum transferred to a particle ejected from a moving space ship will be proportional to the relative ejection velocity. Let us use a similar logic for the emission of photon particles from a moving source. To begin with, let us assume that the light source is at rest in the universal reference frame X. Let E_0 be the energy and p_0 be the momentum transferred to an emitted photon under certain electron transition. Now consider the light source to be moving with uniform velocity **u** in the reference frame X. Let a photon be emitted under similar electron transition as before such that the emitted photon, with energy E, moves with velocity **c** in reference frame X. The magnitude of momentum p and energy E transferred to the photon by the light source in motion will be proportional to the relative emission velocity $\mathbf{V_r}$ such that:

$$\mathbf{V_r} = \mathbf{c\text{-}u} \quad (12.15)$$

And

$$p = p_0.|\mathbf{V_r}|/c = p_0.\,|\,\mathbf{c\text{-}u}\,|\,/c \quad (12.16)$$

$$E = E_0.\,|\mathbf{V_r}|/c = E_0.\,|\,\mathbf{c\text{-}u}\,|\,/c \quad (12.17)$$

In essence, this is the new hypothesis regarding emission of photons from a moving source. Using the relations $E_0 = h.f_0$ and $p_0 = h/L_0$ we get the corresponding values for wave length L and frequency f from equations (12.16) and (12.17) as,

$$L = L_0.c/|\mathbf{V_r}| = L_0.\ c\ /|\ \mathbf{c\text{-}u}\ | \qquad (12.18)$$

$$f = f_0.\ |\mathbf{V_r}|/c = f_0.\ |\ \mathbf{c\text{-}u}\ |\ /c \qquad (12.19)$$

Therefore, *we finally propose a new hypothesis as per which the energy of a photon emitted from a moving light source is proportional to the magnitude of relative emission velocity* V_r or more correctly V_r/c. Now, we show that the null result of the MM experiment is fully explained by using the proposed hypothesis. This constitutes the main justification for its validity, apart from the well known thermal broadening of emission spectral lines.

12.5 Explanation of the Null Result of MM Experiment

In the light of the proposed new hypothesis, let us reconsider the case II above. Here, Platform M moves along OX at uniform velocity **v** and initial beam of photons is emitted from the light source in the OX direction with frequency f_x. The magnitude of relative emission velocity of primary beam of photons is (c-v). From equation (12.19),

$$f_x = f_0.\ (1\text{-}v/c) \qquad (12.20)$$

After the first reflection the relative emission velocity V_y and the frequency f_y of the transverse beam of photons will be given by:

$$V_y = \sqrt{c^2 - v^2} \qquad (12.21)$$

$$f_y = f_0\sqrt{1 - \frac{v^2}{c^2}} \qquad (12.22)$$

The number of wave lengths covered on the path length $A_0Y_1A_3$ of the transverse beam will now be given (using equations (12.4) and (12.22)) by modified equation (12.6) as,

$$N_y = f_y.t_3 = f_0\sqrt{1 - v^2/c^2}.\frac{2D}{c}.\frac{1}{\sqrt{1 - v^2/c^2}} = 2D/L_0 \qquad (12.6A)$$

And $\qquad \phi_y = 2\pi.N_y = 4\pi D/L_0 \qquad$ (12.7A)

After the first reflection, the relative emission velocity V_{-x} and the frequency f_{-x} of the return axial beam of photons will be given by,

$$V_{-x} = c+v \tag{12.23}$$

$$f_{-x} = f_0.(1+v/c) \tag{12.24}$$

After the second reflection of the axial beam at the first mirror A_4 its frequency will change to f_y in common with that of the transverse beam. Therefore, the number of wave lengths covered on the path length $A_0X_2A_4$ of the axial beam will now be given (using equations (12.8), (12.9), (12.10), (12.20)and (12.24)) by modified equation (12.11) as,

$$N_x = f_x.t_2 + f_{-x} . (t_4\text{-}t_2)$$

$$= f_0. (1\text{-}v/c). D/(c\text{-}v) + f_0.(1+v/c) . D/(c + v)$$

$$= 2D/L_0 \tag{12.11A}$$

And $\quad \phi_x = 2\pi.N_x = 4\pi D/L_0 \qquad$ (12.12A)

Hence, as seen from equations (12.6A), (12.7A), (12.11A) and (12.12A) the number of wave lengths covered on the path lengths of both the axial and transverse beams is the same and is independent of the velocity v. As such, both the transverse and axial return beams arrive at the receiver interferometer in phase and no fringe shift is expected on rotation of the platform M. The null result of the MM experiment is therefore fully in conformity with the proposed new hypothesis regarding emission of photons from a moving source.

Therefore, from the satisfactory explanation of the null result of MM experiment, we may conclude that the relative emission velocity of photons from a moving source is dependent on the relative direction of emission in comparison with the direction of motion of the source. After emission from the source, the photons have to propagate at constant speed c in the universal reference frame (vacuum) regardless of the motion of the source. As per the new hypothesis, the energy E and frequency f of the emitted photon will be proportional to the relative emission velocity (**c-v**).

Specifically, let us consider an observer O to be at rest in the universal reference frame and let a source S approach the observer with velocity **v.** The frequency of photons emitted by S in the direction of its velocity vector **v** will therefore be reduced by a factor (c-v)/c. On the other hand the frequency of photons emitted by S in the direction opposite to its velocity vector **v** will be increased by a factor (c+v)/c. Hence the observer O will find the frequency of photons received from an approaching source S to be reduced by a factor of (c-v)/c and that of photons received from a receding source S to be increased by a factor of (c+v)/c.

Such an effect of the motion of light source on the frequency of emitted light is already well known in the temperature dependent broadening of spectral lines. But this is generally attributed to the Doppler effect. However, the frequency change proposed in the new hypothesis is opposite to that predicted by the Doppler effect. That is probably due to the fact that the emission of a light photon is not a continuous but a nearly spontaneous process.

It is important that this new hypothesis be critically examined and tested before formally accepting it as valid. Implications of the new hypothesis could be drastic in cosmology since the observed red shift of distant stars and galaxies will indicate a contracting universe instead of the currently held view of an expanding universe! Moreover the proposed hypothesis is expected to be applicable for the photons emitted from electron transitions in atoms and molecules but not for continuous EM waves radiated from RF circuits and antennas.

13

The Standard Model of Particle Physics

13.1 Elementary Particles

The notion of elementary particles has been significantly revised during the last century. An elementary particle is supposed to be a particle without any substructure; that is, it is not made up of smaller particles. An elementary particle with no substructure, is supposed to be one of the basic building blocks of the universe from which all other particles are made. In the Standard Model of particle physics - quarks, leptons, and gauge bosons are considered to be elementary particles. Earlier, leptons, mesons and baryons such as the proton and neutron, were regarded as elementary particles.

Still, however, the notion of elementary particles embodies an axiomatic assumption regarding their spatial extension, mass, charge and interaction properties. Elementary particles are assumed to be point-like in size, assumed to possess the property of mass, assumed to contain an attribute of charge - positive, negative or zero. Further, elementary particles are assumed to be endowed with special interaction characteristics. All other composite particles of matter are supposed to consist of two or more elementary particles bound together by specific binding interactions.

It is significant to note that the current notion of elementary particles does not accommodate the possibility that these particles could have spatial extension and structure without containing any smaller particles as constituents. Further, the physical properties of mass, charge etc. are supposed to be the fundamental attributes that are somehow carried or possessed by the elementary particles, which cannot be derived from their inherent structure, if any. Of course, it is true that as per the current state of understanding in the field of particle physics, we do not know anything about the shape, size and internal structure of any of the elementary particles. Therefore, such a possibility regarding definite space-time structures or the dynamic structures of elementary particles in the space continuum, needs to be explored.

Detection of Elementary Particles. Most of the particle physics experiments consist of accelerating certain known particles to high velocities, colliding them with equally well known target particles, detecting and studying the properties of various particles and radiations produced during such collisions. The rapid progress in elementary particle physics during the last century can be attributed to a parallel progress in technology and instrumentation. This includes tremendous advancements in particle accelerators, particle beam controls and particle separators that constitute key elements in most of the particle physics experiments. With the use of advanced particle detectors, the creation, flight trajectories and decay characteristics of most elementary particles can be observed. Through the application of current sophisticated techniques, - the mass, charge, kinetic energy, momentum, life time and interaction characteristics of all elementary particles can be determined with a high degree of accuracy.

However, the study of elementary particles through high energy collision / scattering experiments cannot provide sufficient information to determine the detailed space-time structures of the elementary particles involved. Even the most exotic properties of elementary particles such as intrinsic spin, iso-spin, strangeness, charm etc., cannot provide any information on the characteristic space-time structure, like geometrical shape and size, mass and energy density distribution etc. of such elementary particles.

Grouping of Particle Properties. Various stable and unstable elementary particles detected so far, have been grouped into different categories depending on their charge, intrinsic spin, interaction characteristics, mass and decay profiles. At first level all elementary particles can be categorized into either bosons or fermions depending on their spin. In quantum mechanics, the spin of a particle is related to the statistics obeyed by that particle. Spin is the intrinsic angular momentum of a particle and all particles with integer spin are known as bosons and those with half-integer spin are known as fermions. As per the standard model of particle physics, fermions are divided into groups of quarks and leptons. Quarks are further subdivided into up, down, charm, strange, top, and bottom quarks. Leptons are further subdivided into the electron and electron neutrino, muon and muon neutrino, tau and tau neutrino. Bosons are divided into the photon, gluon, W and Z bosons and Higgs boson.

Whereas the electron, positron and muon type particles take part in electromagnetic interaction, the hadrons consisting of mesons, baryons and hyperons take part in strong interaction. Further, the electron and electron neutrino are considered first generation lepton particles and the muon, muon neutrino are considered second generation leptons. These generations of particles are interrelated through their decay profiles. The hadrons are supposed to be made up of various quarks. Primarily, quarks are supposed to take part in all strong interactions, even though, these are assumed to be incapable of independent existence.

13.2 Standard Model of Elementary Particles

In physics, the Standard Model of particle physics is currently the best description or representation of all experimental data. Apart from categorizing a large number of stable and unstable elementary particles into various groups of similar characteristics, mathematical group theory has been extensively used to depict their mutual interactions, transformations and dynamic behavior. In fact, the concept of a group is a central concept of modern abstract algebra and the theory of abstract mathematical representations. Such a mathematical representation of almost all particles observed in high energy scattering and collision experiments has practically covered all facets of their physical properties except for their shape, size and internal structure. A mathematical group representation theory studies properties of abstract groups via their representations as linear transformations of vector spaces.

In quantum mechanics, the state of the world is represented by a vector in complex vector space H, the Hilbert space. There is a unitary representation of any symmetry group G of the theory on the complex vector space. Gell-Mann discovered that strongly interacting particles could be organized into irreducible representations of SU(3) group, with the iso-spin SU(2) as a subgroup. Later, it became clear that the SU(3) symmetry Gell-Mann had found, was due to the existence of three relatively light particles called quarks, which fit into the fundamental three dimensional representation of SU(3).

The gauge theory for the SU(3) group is called Quantum Chromodynamics (QCD). The formalism for QCD and electroweak interaction together is known as the Standard Model, which unifies the *mathematical description* or representation of electromagnetism, weak interactions and strong interactions in the language of gauge theory. The Standard Model of particle physics contains 12 flavors of elementary fermions and their corresponding antiparticles, as well as elementary bosons that are supposed to mediate the forces. The construction of the standard model proceeds by first postulating a set of symmetries of the system, and then by writing down the most general re-normalizable Lagrangian from its particle or field content that observes these symmetries.

The Standard Model, based on the gauge group SU(3) x SU(2) x U(1), is one of the great successes of the gauge theory. It can describe or provide *mathematical representations* for all known fundamental forces except gravity. However, the Standard Model is certainly not the final physical theory of particle interactions. It involves 18 unknown parameters and cannot explain the origin of the quark masses or the various coupling constants. The model is rather unwieldy and inelegant. Nevertheless, it is re-normalizable and it can explain a vast number of results from all areas of particle physics. In that sense the Standard Model represents an excellent empirical mathematical model of particle physics. However, the Standard Model contains too many free parameters. With so much arbitrariness, the Standard Model should be considered only as an empirical approximation to the true theory of subatomic particles, i.e., *it is an effective empirical model which needs to be explained by a more fundamental theory.*

13.3 Exchange Theory of Interactions

There are four fundamental interactions among particles of matter, namely gravitational, electromagnetic, weak and strong interactions. As per the currently accepted theories of physics, all interactions among interacting particles are mediated by the exchange of special force mediating particles called bosons. In particular, as per the Standard Model, electromagnetic interaction among charged particles is supposed to be mediated through the exchange of photons; weak interaction in radioactive decay is supposed to be mediated through the exchange of W and Z bosons and strong interaction among quarks (the assumed building blocks of hadrons) is supposed to be mediated through the exchange of gauge bosons - called gluons. An elaborate mathematical representation of particle interactions, through force mediating bosons, has been developed in the group theory of the Standard Model.

However, the exchange theory of interaction is based on a set of ad-hoc assumptions regarding the emission and absorption of mediating (virtual) bosons from point like interacting particles. There is no supporting physical mechanism that could justify the exchange theory of interaction. Specifically, following points need to be highlighted in this regard:

(a) When two interacting particles A and B are required to exchange appropriate bosons for mediating the interaction, they will have to first exchange information regarding each others position, momentum, charge, mass etc. for determining the properties and direction of the boson to be emitted. There is no physical mechanism envisaged in the theory for exchange of such information.

(b) If prior exchange of required information between the interacting particles is not possible, then the particles will have to keep emitting infinitely large number of (virtual) bosons in all possible directions at all times. There is no physical mechanism for doing so. This constraint is overcome through ad-hoc assumptions.

(c) When we consider simultaneous interactions among a large number of different elementary particles, the need for prior exchange of relevant information among interacting particles becomes unavoidable to ensure specific interactions through exchange of appropriate (virtual) bosons among different pairs of particles. There is no physical mechanism available for such intelligent interactions.

13.4 Inadequacy of the Standard Model

It is true that the Standard Model represents an excellent empirical mathematical model, which has been fine tuned with most of the experimental observations in high energy particle physics. However, this model cannot qualify for a valid physical theory essentially because it is founded on too many ad-hoc assumptions and ad-hoc parameters. The most glaring of these ad-hoc assumptions is the exchange theory of interactions discussed above. Once we invalidate the assumption of the exchange theory of interaction, the whole structure of the Standard Model of particle physics collapses on itself.

Even though the Standard Model provides an excellent mathematical representation of all particle properties observed in high energy scattering and collision experiments, it has practically failed to provide adequate information on the characteristic shape, size and internal structure of any of the elementary particles. It is therefore quite obvious that even if we retain the Standard Model as an excellent empirical mathematical model, we cannot afford to abandon the search for a more appropriate theory of elementary particles and their interactions.

PART II

Space Dynamics to Particle Structures

14

Dynamics of the Space Continuum

14.1 Space-time Distortions or Space Dynamics

An abstract mathematical notion of spacetime curvature had dominated fundamental physics during the last century to such an extent that most scientists now perceive it as a common physical notion. Mathematically, however, the term '*spacetime curvature*' implies a non-zero value of the Riemann tensor computed from the metric coefficients of the 4D spacetime manifold. In any gravitation free region of space, the metric of 4D spacetime manifold may be represented by $g_{ij}(x)$ such that the corresponding Riemann tensor is zero and an infinitesimal separation distance ds between two neighborhood points $P(x^i)$ and $Q(x^i+dx^i)$ is given by:

$$(ds)^2 = g_{ij}\, dx^i\, dx^j \tag{14.1}$$

However, in the same region of space under the influence of a gravitational field, the metric of 4D spacetime manifold will be represented by $h_{ij}(x)$ (as per EFE) such that the corresponding Riemann tensor is non-zero and an infinitesimal separation distance ds' between the neighborhood points P' and Q' is given by:

$$(ds')^2 = h_{ij}\, dx^i\, dx^j \tag{14.2}$$

To distinguish between a rigid and a deformable continuum of space points, let P be any point in this continuum and P_1, P_2, ... P_n be n points in the neighborhood of P. Let ds_1 be the separation distance between points P and P_1, ds_2 the distance between P and P_2, ..., ds_n the distance between P and P_n. If these separation distances ds_1, ds_2, ... ds_n from P to all of its neighborhood points, remain constant under all circumstances, then the continuum under consideration can be regarded as rigid. If under certain circumstances, these separation distances change to say ds_1', ds_2', ... ds_n' then the continuum under consideration will be regarded as deformable. Since, as per General Theory of Relativity (GR), the separation distance ds between two neighborhood points of the spacetime continuum does change under the influence of gravitational field, the spacetime continuum is assumed to be deformable in GR. Considering only the spatial components of the metric tensor, it can be shown that the separation distance ds between two neighborhood points of the space continuum also changes under the influence of gravitational field (as per GR). Therefore the space continuum is also assumed to be deformable in GR.

Obviously, whenever the separation distance between neighboring points P and Q in the space continuum, changes from ds to ds' it implies a relative shift in the original positions of P and Q to the changed positions say P' and Q' such that arc element P'Q'=ds'. This relative shift in positions of P and Q to the changed positions P' and Q' may be referred as the relative displacement of these points. Specifically, the vector PP' may be defined as the displacement vector U and the corresponding displacement of Q to Q' will then be represented by the incremented displacement vector U+δU (Fig.14.1). Hence, whenever the separation distance ds between two neighboring points P and Q changes to ds' as given by equations (14.1) and (14.2), the associated displacement vector field **U** will be defined at all points P of the space continuum.

Figure 14.1

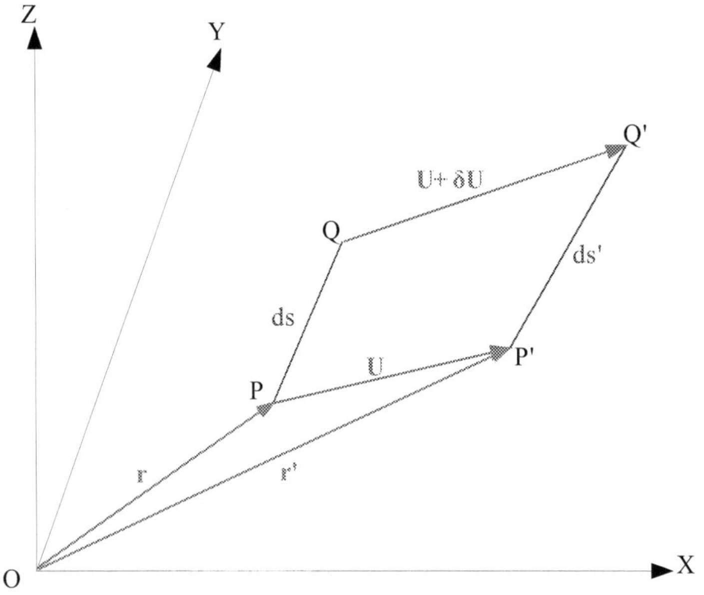

The deformation of the continuum can be said to be fully determined when the displacement of every point P in the continuum is known or uniquely determined. The existence of displacement vector **U** at every point P as a function of position coordinates of P, will constitute a displacement vector field **U** in the continuum. The displacement vector from point P(**r**) to P'(**r'**) is given by the relation,

$$\begin{aligned}\mathbf{U} &= \mathbf{r'} - \mathbf{r} \\ &= u^i \, \mathbf{a_i}\end{aligned} \qquad (14.3)$$

where, u^i are the components of vector **U** and **r**, **r'** are the position vectors of P and P'.

In general, the displacement vector field **U** in the space continuum will be a function of space coordinates and time. As such, the time dependent deformations of physical space, that could be represented through a time dependent displacement vector field **U**, may be described as space-time distortions. The space and time derivatives of **U** will correspond to a strain tensor field in the space continuum. Since the above referred time-dependent deformations in the space continuum are reversible, the physical space continuum can be assumed to be elastic in nature. Hence, logically the displacement vector and strain tensor fields will also be accompanied by the corresponding stress tensor field in the physical space continuum. The study of space-time distortions through the detailed study of corresponding displacement vector, strain tensor and stress tensor fields in the space continuum may be termed as space dynamics. A detailed analytical study of space dynamics can provide us a valuable insight into the fundamental structures of various elementary particles and their interactions.

14.2 Localized Regions under Dynamic Strain

Under the proposed viewpoint, the electromagnetic fields as well as all elementary particles could be viewed as space-time distortions or dynamic deformations in the space continuum or equivalently the dynamic stress strain fields in a bounded region of the physical space continuum. That is, these deformations or the corresponding stress strain fields will constitute a structure of these particles and cover a finite bounded region of physical space. Therefore, we need to shed the current notion of elementary particles being point or point-like particles with characteristic properties. Instead, we need to derive the characteristic properties from the structure and the interactions of these particles.

There are two reasons why we need to move beyond the conventional view-point, even if it is a difficult and arduous task. Firstly, we need to understand the fundamental nature of electromagnetic fields as well as all elementary particles with such clarity that we should be able to mentally visualize the same, just as we do most other physical entities. Secondly, we need to understand that the universally accepted notion of space curvature actually implies deformation of the space continuum. However, in scientific literature, the notion of deformation of space is not as popular and as well accepted as the notion of space curvature. When the Euclidean metric of space continuum is changed to Riemannian metric with non-zero value of corresponding Riemann tensor, it automatically implies deformation of the space continuum. A detailed analysis of the incompatibility of space curvature induced deformations leads to the invalidity of General Theory of Relativity.

Electromagnetic Field as Dynamic Strain Field. Electric and magnetic fields could be viewed as dynamic deformations in the space continuum with physical properties of ε_0, μ_0 and c (with $\varepsilon_0.\mu_0=1/c^2$). Let **U** be a time dependent displacement vector in the space continuum such that it satisfies the dynamic equilibrium equation or the vector wave equation.

$$\nabla^2\mathbf{U} = (1/c^2)\ \partial^2\mathbf{U}/\partial t^2 \tag{14.4}$$

A solution of equation (14.4) for **U** that satisfies the essential boundary conditions, will represent a transverse wave field if $\nabla.\mathbf{U}=0$. Further, we may identify **U** with the conventional electric and magnetic fields **E** & **B** in free space through the following identities:

$$\mathbf{D} = -\ (1/c).\ \partial\mathbf{U}/\partial t \tag{14.5}$$

$$\mathbf{E} = (1/\varepsilon_0).\mathbf{D} = -(1/\varepsilon_0).(1/c).\ \partial\mathbf{U}/\partial t \tag{14.6}$$

$$\mathbf{H} = c.(\nabla \times \mathbf{U}) \tag{14.7}$$

$$\mathbf{B} = \mu_0.\mathbf{H} = (1/c).(1/\varepsilon_0).(\nabla \times \mathbf{U}) \tag{14.8}$$

From equations (14.6) and (14.8),

$$(\nabla \times \mathbf{E}) = -\ \partial\mathbf{B}/\partial\ t \tag{14.9}$$

$$\begin{aligned}(\nabla \times \mathbf{H}) &= c.(\nabla \times (\nabla \times \mathbf{U}))\\ &= c.\ (0-\nabla^2\mathbf{U}) \qquad \text{(using } \nabla.\mathbf{U}=0)\\ &= -(1/c).\partial^2\mathbf{U}/\partial\ t^2 \quad \text{(from equation 14.4)}\\ &= \partial\mathbf{D}/\partial\ t\end{aligned} \tag{14.10}$$

The displacement vector field **U** will thus satisfy all the electromagnetic field equations that are satisfied by E & B in free space. Thus the displacement vector field **U** may also be viewed as a unification parameter for the electric and magnetic fields.

Solution of Dynamic Equilibrium Equation. Equation (14.4) could be an excellent starting point to develop some insight into the highly complex phenomenon of dynamic deformations or stress strain fields in the space continuum. We may attempt to solve equation (14.4) in terms of components of displacement vector **U** in any convenient coordinate system subject to appropriate boundary and stability conditions. For example, in Cartesian coordinates, we have to solve equation (14.4) for displacement vector components u_x, u_y and u_z as functions of x, y, z and t. Such a solution can lead us to the space-time structure of the photon wave packet. In spherical polar coordinates, we may solve for the physical components u_r, u_θ and u_ϕ of vector **U,** as functions of r, θ, ϕ and t. A spherically symmetric solution of equation (14.4), depicting standing strain wave oscillations and satisfying appropriate boundary and stability conditions, can lead us to the space-time structure of the electron (/positron) type particles.

It is very important to clarify two points regarding space-time distortions or space dynamics of stress strain fields. Firstly, all space-time distortions or dynamic deformations of the space continuum can be regarded as strained states of the continuum. Secondly, all such dynamic deformations of the space continuum will always be associated with corresponding energy density of the strained state. The space-time derivatives of the displacement vector **U** will yield the strain components like $\partial u_x/\partial x$ or $\partial u_r/\partial r$. In a particular deformed or strained region of the space continuum, the strain energy density is given by,

$$W = (1/2\varepsilon_0)*[\text{sum of squares of all strain components}] \qquad (14.11)$$

As mentioned earlier, all permissible solutions of equation (14.4), in a specified deformed or stressed region of the space continuum, will have to satisfy appropriate boundary and stability conditions. The most important boundary and stability conditions that are generally applicable, are,

(a) All displacement vector components u_i must be finite and continuous within the specified region and must vanish at the boundaries of the specified deformed or strained region of the space continuum.

(b) All strain components $u_{i,j}$ must be finite and continuous within the specified region of the space continuum.

(c) The total strain energy within the specified region, obtained by integration of the energy density W over the whole region, must be finite and constant or invariant with time.

Localized Strain Bubbles. Logically, for any observably finite and distinct region of strained space continuum, we should be able to find appropriate solutions of equation (14.4) which satisfy the above mentioned boundary and stability conditions. But that is an extremely difficult exercise. Putting it another way around, if the solutions of equation (14.4) in a finite region of strained space continuum satisfy the above mentioned boundary and stability conditions, then that region of the space continuum will be found to be an observably distinct entity.

Such distinct regions of the space continuum with finite, stable, total energy content, may be called Strain Bubbles, which will behave like elementary particles. Here it may be pointed out that none of the permissible solutions of equation (14.4) is a static or time independent solution.

14.3 Strain Bubble representing Electron Structure

One particular spherically symmetric solution of equation (14.4), in terms of displacement vector components u_r and u_ϕ, shows the electron structure as consisting of a central core of about 1.61 fm (10^{-15} m) radius containing a standing strain wave field and surrounded by a radial phase wave electrostatic field with decaying amplitude. The radial wave field for electron is given by $f(r).e^{iK(r+ct)}$ and that for positron by $f(r).e^{iK(r-ct)}$. Here, K represents the wave number of the radial wave field and could be of the order of 10^{15} m^{-1}. The amplitude factor f(r) is proportional to 1/r. The concept of charge is related to the direction of propagation, intensity and interaction characteristics of the radial wave field.

This picture of the electron is drastically different from the conventional point mass and point charge notion generally taken for granted. In this ***core-field*** picture of an electron (or positron), the mass energy is characteristically distributed in space and its charge property is represented by the interaction characteristic of its wave field. This radially decaying wave field replaces the notion of virtual photons. About 65 percent of the total mass energy of the electron (positron) is contained in the central core region and the remaining 35 percent is distributed in its wave field. The characteristic frequency of oscillations of the standing wave field of the electron/positron core is of the order of 8×10^{22} Hz.

When two opposite charges interact, their electrostatic wave fields get superposed thereby reducing the amplitude of the resultant wave field and reducing the combined field energy of the interacting charges. This reduction in the combined field energy amounts to a net release of a portion of their field energy, called negative interaction energy, or the potential energy. The interaction energy thus released could either get transferred to kinetic energy of the interacting charges, or get used up in creation of a photon or some other transient elementary particle. If the released interaction energy is given out, or gets extracted from the system then the interacting charge particles are said to get bound together and the interaction energy extracted from the system is termed as their binding energy or the so called negative 'total energy' E of the system.

When two similar charges interact, their electrostatic wave fields get superposed, thereby increasing the amplitude of the resultant wave field. This superposition will lead to the corresponding increase in the combined field energy of the interacting charges, thereby leading to positive interaction energy or the potential energy. The Coulomb interaction between two charged particles is essentially the interaction between their radially decaying wave fields and is strictly valid for separation distances greater than 3.2 fm.

14.4 Photon Wave Packet as a Propagating Strain Bubble

As an electromagnetic wave packet, the electric and magnetic field vectors within a photon must satisfy Maxwell's vector wave equation and must account for a finite energy content of the packet. In terms of displacement vector field representation for the photon, the vector field U_p must satisfy the above conditions and must vanish at the boundaries of the wave packet. Considering a Cartesian coordinate system XYZ, let a photon wave packet be propagating along X-axis at constant velocity c. Then as per a tentative model, the displacement vector components (u_x, u_y and u_z) of a photon wave packet, may be given as,

(a) $u_z = 0$

(b) u_x and u_y to be exponentially decaying sinusoidal functions of $-|x-ct|$ and y, z such that they satisfy equation (14.4) as well as the boundary and stability conditions.

(c) The amplitude of u_x and u_y must be maximum at the center of the packet.

From these displacement vector components u_x and u_y, we can compute the strain tensor components and hence the total energy content in the wave packet. Such detailed computations show that the total energy content within the photon wave packet comes out to be proportional to its frequency only if the displacement components u_x and u_y themselves are taken to be proportional to the wave number k (where $k=2\pi.f/c$). It can thus be seen from the spatial extension of displacement components u_x and u_y that the photon wave packet is essentially just a small sinusoidal pulse of displacement vector field U_p with exponentially decaying amplitude. In other words, the photon may be viewed as a sinusoidal pulse of electromagnetic field E_p & B_p with exponentially decaying amplitude and 'significant' spatial extension of just about one wave length in all directions.

The Photon Interaction. Here it is appropriate to point out that similar to the computation of Coulomb Interaction, the mutual interaction of two or more photons separated by distance 'b' along any Cartesian coordinate axis, can also be easily computed. This is done by superposition of the strain tensor components of two interacting photons separated by distance b along any coordinate axis and referred to a common Cartesian coordinate system. The strain energy of the superposed or combined field can then be easily computed. The computation results show that the interaction energy for two photons of the same frequency, separated by a distance 'b', depends on functions of the type $2h.f.e^{-k.b}.cos(k.b)$. That is, any two photons of the same frequency f, will tend to get mutually coupled at a certain optimum separation of the order of an 'odd number of half wave lengths'. Their interaction energy will change from negative to positive if their separation along any coordinate axis is changed by about one half wave length, resulting in their mutual repulsion. This may account for the conventional interference and dispersion effects encountered in a stream of photons of the same frequency. Further, this characteristic interaction between adjoining photons of the same frequency, also transforms a continuous stream of photons into an effectively continuous wave beam of light.

Case for the Study of Space Dynamics. The foregoing discussions have shown that a detailed analytical study of dynamic deformations in the physical space continuum, through the time dependent displacement vector fields, can yield valuable insight into the structure of elementary particles and fields. The study of time dependent displacement vector field provides a more fundamental level of investigation into the workings of Nature, in comparison to the fields currently employed for the purpose. Apart from providing unification between electric and magnetic fields, the displacement vector field throws up a mechanism to demarcate the physical boundaries of elementary particles. Therefore, we must exploit the vast potential of space dynamics to explore the fundamental structure of all elementary particles and their interactions.

15

Equilibrium Equations in Elastic Space Continuum

15.1 Introduction

In the 19th century Physics, light waves were regarded as undulations in an all pervading elastic medium called aether. The successful explanation of diffraction and interference phenomenon in terms of aether waves made the notion of the aether so familiar that its existence was taken for granted. However, in order to accommodate the notion of aether in the framework of physical universe as known at that time, some self contradicting properties had to be ascribed to this medium. For supporting light waves, aether was required to behave like an elastic solid. But to enable the motion of material bodies through it without any resistance, it had to behave like a thin, ideal, non-viscous fluid. Maxwell's development of the electromagnetic theory of light, rendered the aether medium superfluous as the electromagnetic (EM) field was granted an independent status, capable of independent existence and propagation in free space in accordance with Maxwell's equations. Further, consequent to the null result of MMX, the aether got discarded from the 20th century Physics so thoroughly that its non-existence is now taken for granted.

Of course, granting of independent status to the EM field was not sufficient by itself, we had to ascribe the characteristic properties of permittivity 'ε_0' and permeability 'μ_0' to empty space i.e. 'nothingness', which again appears self contradictory. Propagation of independent EM field through empty space, at a constant velocity c, also depended upon the magnitude of characteristic parameters ε_0 and μ_0 ascribed to empty space. If the characteristic parameters ε_0 and μ_0 are associated with an 'elastic continuum' pervading the entire space, we could view the EM waves, with energy stored in their oscillating electric and magnetic fields, as propagating through this continuum. Hence, *the transportation of energy across physical space could be viewed as a propagation process of specific type of waves through the elastic space continuum.*

However, as the next most formidable step, it is extremely difficult to imagine the transportation of matter as a sort of propagation process through the elastic space continuum, even though 20th century Physics has shown the equivalence and inter-convertibility between matter and energy. It is one thing to imagine elementary particles as some sort of packets of energy entrapped in a characteristic wave formation in the space continuum, but quite another to imagine the transportation of clusters of such particles (material bodies) as a sort of 'propagation' process through the continuum. Yet, this most formidable step is also the most crucial one that is necessary to divest the self contradictory property of 'thin, ideal fluid' from the notion of an elastic aether or space continuum. Therefore, it seems likely that all the EM phenomena and all energy entrapping and transportation processes that we believe to be occurring in empty space, are in fact occurring in the elastic space continuum with the characteristic properties of permittivity ε_0 and permeability μ_0 or an elasticity constant $1/\varepsilon_0$ and inertial constant μ_0. This revised notion of aether no longer requires it to be 'thin, ideal fluid' to allow free unrestricted motion of matter through it, since matter is no longer considered an independent entity separate from the aether. Therefore, to distinguish this revised notion from the old aether medium of 19th century, we may simply call it the Elastic Space Continuum with associated characteristic parameters of elastic constant $1/\varepsilon_0$ and inertial constant μ_0. We may consider that we are just reinterpreting our familiar concept of space continuum with associated parameters ε_0 and μ_0, as the 'Elastic Space Continuum' with the associated parameters $1/\varepsilon_0$ and μ_0.

15.2 General Equations of Elasticity in an Elastic Continuum

Displacement Vector Field U. Let us consider an isotropic elastic continuum pervading the entire space. Initially, let all the physical points of this continuum be represented by the corresponding geometrical points of our familiar three dimensional space referred to a suitable orthogonal coordinate system. In a conventional Cartesian coordinate system, let the x, y, z coordinates be represented by x^1, x^2, and x^3 respectively and the corresponding unit vectors **i, j, k** be represented by $\mathbf{e_1}, \mathbf{e_2}, \mathbf{e_3}$. If O is the origin, the position vector of any point $P(x^1,x^2,x^3)$ or simply $P(x^i)$ will be given by,

$$\mathbf{OP} = \mathbf{e_1}\, x^1 + \mathbf{e_2}\, x^2 + \mathbf{e_3}\, x^3 = \mathbf{e_i}\, x^i \quad \text{(summation over i from 1 to 3)}$$

With the passage of time, physical points of the continuum may undergo infinitesimal displacements **U,** leading to time dependent infinitesimal deformations in the continuum. The infinitesimal displacement at any point $P(x^i)$ may be represented by a displacement vector **U** as a function of the coordinates of P as well as 't'.

$$\mathbf{U}(x^i,t) = \mathbf{e_1} u^1(x^i,t) + \mathbf{e_2}\, u^2(x^i,t) + \mathbf{e_3}\, u^3(x^i,t) = \mathbf{e_j}\, u^j(x^i,t) \qquad (15.1)$$

If this displacement vector **U** is finite and continuous within a region of space V, then a displacement vector field $\mathbf{U}(x^i,t)$ may be said to be defined over this region of space. Specifically, this displacement vector field **U**, represented by its components u^i, may be a periodic function or a combination of periodic functions of coordinates x^i and t within the field region V and may be zero at the boundaries & outside this region. Obtaining specific solutions for the displacement vector field $\mathbf{U}(x^i,t)$, under specified initial and boundary conditions, will be our major objective in the study of Elastic Space Continuum.

Representation of Strain S. The displacement vector field $\mathbf{U}(x^i,t)$ will also represent an infinitesimal deformation field in the Elastic Continuum. The infinitesimal deformation at any point $P(x^i,t)$ is best quantified through the components of a strain tensor **S** as follows. The infinitesimal deformation or change of an arbitrary small vector $A^i(x^1,x^2,x^3,t)$ at the point $P(x^1,x^2,x^3)$ will be given by an infinitesimal affine transformation of the neighborhood of the point in question as,

$$\delta A^i = (\partial u^i/\partial x^j)\, A^j = u^i_{,j}\, A^j \quad \text{(summation over j)} \tag{15.2}$$

Here, the quantities $u^i_{,j}$ which are the covariant derivatives of the displacement vector field u^i with respect to the x^j represent the components of strain tensor **S** such that,

$$S^i_j = u^i_{,j} \tag{15.3}$$

These components obviously represent only the spatial strain components. Since the displacement vector components u^i in general will be functions of space and time coordinates, the partial derivatives of u^i with respect to time t (more correctly ct) that is, $(1/c).\partial u^i/\partial t$ will constitute temporal strain components. In accordance with the notions of 4-D space-time manifold, time can be regarded as a fourth dimension coordinate at right angles or in quadrature to the three space coordinates. Similarly, the temporal strain component $S^i_t=(1/c).\partial u^i/\partial t$ can also be regarded as being in quadrature to the corresponding spatial strain components $S^i_j = u^i_{,j}$. However, if the fourth dimensional coordinate is taken as $x^4 = \boldsymbol{i}ct$, where $\boldsymbol{i}$ is the complex number $\sqrt{-1}$ then corresponding to the displacement vector components u^i the temporal strain components can be written as,

$$S^i_4 = u^i_{,4} = \partial u^i/\partial x^4 = (1/\boldsymbol{i}\, c).\partial u^i/\partial t \tag{15.4}$$

Thus, corresponding to three components of the displacement vector $u^i(x^1,x^2,x^3,t)$, there will be nine spatial strain components and three temporal strain components, all of which will be functions of the space and time coordinates.

In contrast to the Elastic Space Continuum considered above where no rigid body motion is possible, the infinitesimal deformation in elastic material media is generally split into pure deformations and rigid body motions (translations and rotations). For steady state elastic equilibrium in a material media, the spatial strain components e_{ij} representing pure deformation and rotational components ω_{ij} representing rigid body motion, are given by,

$$e_{ij} = (u_{i,j} + u_{j,i})/2 \quad \text{and} \quad \omega_{ij} = (u_{i,j} - u_{j,i})/2$$

However, in the study of the Elastic Space Continuum where rigid body motion is not possible, we shall not use the above mentioned e_{ij} representation for strain components. As discussed in the foregoing, we shall continue to use the total strain tensor components given by,

$$S^i_j = u^i_{,j} \qquad (i \to 1 \text{ to } 3 \ \& \ j \to 1 \text{ to } 4) \qquad (15.5)$$

It may be quite pertinent to mention here that the displacement vector **U** and strain tensor **S** are absolute entities and are invariant under coordinate transformations. Only the magnitude of components u^i and S^i_j is dependent on the reference coordinate system and transform with coordinate transformation. Hence, the analysis of strained state of the Elastic Space Continuum is equally valid in all admissible coordinate systems; even though we generally prefer to use a particular coordinate system for particular problems on the overall considerations of symmetry and boundary conditions.

Representation of Stress T. At any point $P(x^1,x^2,x^3)$ of the Elastic Continuum under infinitesimal deformation, the state of stress is represented by stress tensor **T**, the components τ^i_j of which are defined as follows. With point $P(x^1,x^2,x^3)$ as the center, consider an infinitesimal plane rectangular surface area $\sigma_1 = \delta x^2.\delta x^3$, with its normal parallel to X^1- axis. This infinitesimal area will have two faces. We shall consider that face of σ_1, where its unit normal ν_1 points towards positive X^1-axis, as +ve face and denote it as σ_{+1}. The other face, with normal pointing towards negative X^1-axis, will be considered -ve face and denoted as σ_{-1}. If the net force per unit area acting on σ_{+1} is termed $\mathbf{T_1}$, then it is obvious that the direction of $\mathbf{T_1}$ will not coincide with unit normal ν_1 in general, since this net force represents a resultant of three components. In fact this $\mathbf{T_1}$ vector acting on σ_{+1}, can be decomposed into its components along X^1, X^2 and X^3 coordinate directions as,

$$\mathbf{T_1} = \mathbf{e_1}\tau^1{}_1 + \mathbf{e_2}\tau^2{}_1 + \mathbf{e_3}\tau^3{}_1 = \mathbf{e_i}\tau^i{}_1 \qquad (15.6)$$

In general, for an infinitesimal rectangular plane area σ_{+j} perpendicular to x^j coordinate direction, the net force per unit area $\mathbf{T_j}$ acting on σ_{+j} will be given by

$$\mathbf{T_j} = \mathbf{e_1}\tau^1_{\ j} + \mathbf{e_2}\tau^2_{\ j} + \mathbf{e_3}\tau^3_{\ j} = \mathbf{e_i}\tau^i_{\ j} = \mathbf{e_i}\tau^i_{\ j} \tag{15.7}$$

Here the quantities $\tau^i_{\ j}$ are the components of the stress tensor **T** at point $P(x^1,x^2,x^3)$. The stress components $\tau^i_{\ j}$ in general will be functions of space coordinates (x^1,x^2,x^3) of point P and time t.

The stress components $\tau^i_{\ j}$ are reckoned +ve if the corresponding components of force act in the directions of increasing x^i, when the surface normal is along increasing x^j axis. If on the other hand the surface normal is along -ve x^j axis, then positive values of components $\tau^i_{\ j}$ are associated with forces directed along -ve x^i axis. Hence, for an infinitesimal volume element $\delta V = \delta x^1.\delta x^2.\delta x^3$ taken in the shape of a rectangular parallelepiped, with faces parallel to coordinate planes and point $P(x^1,x^2,x^3)$ as its center, the stress components $\tau^i_{\ j}$ will correspond to forces in opposite directions at the opposite ends of the parallelepiped.

Figure 15.1 A cross section of an infinitesimal parallelepiped under stress.

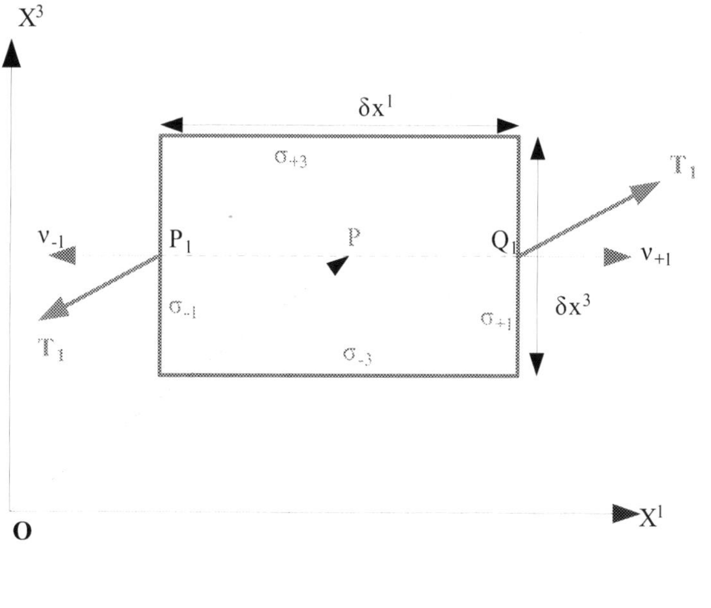

Dynamic Equilibrium Equations. Let us consider an infinitesimal volume element $\delta V = \delta x^1.\delta x^2.\delta x^3$ in the shape of a rectangular parallelepiped, with point $P(x^1,x^2,x^3)$ as its center and faces parallel to coordinate planes.

Of this volume element, let us consider two plane faces σ_{+1} and σ_{-1} perpendicular to the X^1 axis, such that a point $P_1(x^1-½\delta x^1,x^2,x^3)$ is the center of σ_{-1} and point $Q_1(x^1+½\delta x^1,x^2,x^3)$ is the center of σ_{+1} (Fig. 15.1). Then, $P_1Q_1=\delta x^1$; $P_1P=½\delta x^1=PQ_1$ and areas of two opposite plane faces under consideration are $\sigma_{+1}=\delta x^2.\delta x^3=\sigma_{-1}$. At any instant of time t, let us examine total forces acting on faces σ_{+1} and σ_{-1} due to the combined effect of shear and normal stresses acting on these faces.

From equation (15.6), the total force acting on +ve face σ_{+1} is,

$$\sigma_{+1}.\mathbf{T_1}(x^1+½\delta x^1,x^2,x^3) = \delta x^2.\delta x^3[\mathbf{e}_1.\ \tau^1{}_1(x^1+½\delta x^1,x^2,x^3) + \mathbf{e}_2.\ \tau^2{}_1(x^1+½\delta x^1,x^2,x^3) + \mathbf{e}_3.\ \tau^3{}_1(x^1+½\delta x^1,x^2,x^3)]$$

Or $$\sigma_{+1}.\mathbf{T_1}(x^1+½\delta x^1,x^2,x^3) = \delta x^2.\delta x^3[\mathbf{e}_1.\ \tau^1{}_1(x^1-½\delta x^1,x^2,x^3)+ (\partial\tau^1{}_1(P)/\partial x^1).\delta x^1\} + \mathbf{e}_2.\ \tau^2{}_1(x^1-½\delta x^1,x^2,x^3) + (\partial\tau^2{}_1(P)/\partial x^1).\delta x^1\} + \mathbf{e}_3.\ \tau^3{}_1(x^1-½\delta x^1,x^2,x^3)+ (\partial\tau^3{}_1(P)/\partial x^1).\delta x^1\}] \qquad (15.8)$$

And the total force acting on the negative face σ_{-1} is,

$$\sigma_{-1}.T_1(x^1-½\delta x^1,x^2,x^3) = \delta x^2.\delta x^3[\ e_1.\ \tau^1{}_1(x^1-½\delta x^1,x^2,x^3) + e_2.\ \tau^2{}_1(x^1-½\delta x^1,x^2,x^3) + e_3.\ \tau^3{}_1(x^1-½\delta x^1,x^2,x^3)] \qquad (15.9)$$

Therefore, the net resultant force acting on two opposite faces σ_{+1} and σ_{-1} of the parallelepiped is obtained by subtracting equation (15.9) from (15.8) as,

$$\boldsymbol{\delta}\mathbf{T_1}.\ \delta x^2.\delta x^3 = \delta x^2.\delta x^3[\mathbf{e}_1.(\partial\tau^1{}_1(P)/\partial x^1).\delta x^1 + \mathbf{e}_2.(\partial\tau^2{}_1(P)/\partial x^1).\delta x^1 + \mathbf{e}_3.(\partial\tau^3{}_1(P)/\partial x^1).\delta x^1]$$

Similarly, considering the forces on opposite faces σ_{+2}, σ_{-2} and σ_{+3}, σ_{-3} we get the corresponding net resultant forces acting on the parallelepiped as,

$$\boldsymbol{\delta}\mathbf{T_2}.\ \delta x^1.\delta x^3 = \delta x^1.\delta x^3[\mathbf{e}_1.(\partial\tau^1{}_2(P)/\partial x^2).\delta x^2 + \mathbf{e}_2.(\partial\tau^2{}_2(P)/\partial x^2).\delta x^2 + \mathbf{e}_3.(\partial\tau^3{}_2(P)/\partial x^2).\delta x^2]$$

and

$$\boldsymbol{\delta}\mathbf{T_3}.\ \delta x^2.\delta x^1 = \delta x^2.\delta x^1[\mathbf{e}_1.(\partial\tau^1{}_3(P)/\partial x^3).\delta x^3 + \mathbf{e}_2.(\partial\tau^2{}_3(P)/\partial x^3).\delta x^3 + \mathbf{e}_3.(\partial\tau^3{}_3(P)/\partial x^3).\delta x^3]$$

If the body force acting on this infinitesimal volume element δV is $\mathbf{F}(x^1,x^2,x^3)$ *per unit volume*, then in terms of its components along coordinate directions,

$$\mathbf{F}(x^1,x^2,x^3) = \mathbf{e}_1 F^1(x^1,x^2,x^3) + \mathbf{e}_2 F^2(x^1,x^2,x^3) + \mathbf{e}_3 F^3(x^1,x^2,x^3) \qquad (15.10)$$

For equilibrium of the infinitesimal volume element δV, under the action of all resultant surface forces due to spatial stress components and the body forces, we have:

$$\boldsymbol{\delta}\mathbf{T}_1.\ \delta x^2.\delta x^3 + \boldsymbol{\delta}\mathbf{T}_2.\ \delta x^1.\delta x^3 + \boldsymbol{\delta}\mathbf{T}_3.\ \delta x^2.\delta x^1 + \mathbf{F}.\delta x^1.\delta x^2 \delta x^3 = 0 \qquad (15.11)$$

After substituting the values of $\boldsymbol{\delta}\mathbf{T}_1$, $\boldsymbol{\delta}\mathbf{T}_2$, $\boldsymbol{\delta}\mathbf{T}_3$ and $\mathbf{F}$ from the previous equations into equation (15.11), we get,

$$\begin{aligned}
&\delta x^2.\delta x^3.\delta x^1.\ [\mathbf{e}_1.(\partial\tau^1{}_1(P)/\partial x^1) + \mathbf{e}_2.(\partial\tau^2{}_1(P)/\partial x^1) + \mathbf{e}_3.(\partial\tau^3{}_1(P)/\partial x^1)] \\
&+ \delta x^1.\delta x^3.\delta x^2.\ [\mathbf{e}_1.(\partial\tau^1{}_2(P)/\partial x^2) + \mathbf{e}_2.(\partial\tau^2{}_2(P)/\partial x^2) + \mathbf{e}_3.(\partial\tau^3{}_2(P)/\partial x^2)] \\
&+ \delta x^2.\delta x^1.\delta x^3.\ [\mathbf{e}_1.(\partial\tau^1{}_3(P)/\partial x^3) + \mathbf{e}_2.(\partial\tau^2{}_3(P)/\partial x^3) + \mathbf{e}_3.(\partial\tau^3{}_3(P)/\partial x^3)] \\
&+ \delta x^1.\delta x^2.\delta x^3.[\mathbf{e}_1 F^1(x^1,x^2,x^3) + \mathbf{e}_2 F^2(x^1,x^2,x^3) + \mathbf{e}_3 F^3(x^1,x^2,x^3)] = 0 \qquad (15.12)
\end{aligned}$$

However, for the overall resultant of all body and surface forces to vanish, its components along the three coordinate directions should vanish independently. Therefore, the coefficients of $\mathbf{e}_1$, $\mathbf{e}_2$ and $\mathbf{e}_3$ in the above equation (15.12), should vanish separately. Hence,

$$\partial\tau^1{}_1(P)/\partial x^1 + \partial\tau^1{}_2(P)/\partial x^2 + \partial\tau^1{}_3(P)/\partial x^3 + F^1(x^1,x^2,x^3) = 0 \qquad (15.13a)$$

$$\partial\tau^2{}_1(P)/\partial x^1 + \partial\tau^2{}_2(P)/\partial x^2 + \partial\tau^2{}_3(P)/\partial x^3 + F^2(x^1,x^2,x^3) = 0 \qquad (15.13b)$$

$$\partial\tau^3{}_1(P)/\partial x^1 + \partial\tau^3{}_2(P)/\partial x^2 + \partial\tau^3{}_3(P)/\partial x^3 + F^3(x^1,x^2,x^3) = 0 \qquad (15.13c)$$

Or, in tensor notation, the equilibrium equations reduce to a set of three partial differential equations:

$$\tau^1{}_{1,1} + \tau^1{}_{2,2} + \tau^1{}_{3,3} = \tau^1{}_{j,j} = -\,F^1 \qquad (15.14a)$$

$$\tau^2{}_{1,1} + \tau^2{}_{2,2} + \tau^2{}_{3,3} = \tau^2{}_{j,j} = -\,F^2 \qquad (15.14b)$$

$$\tau^3{}_{1,1} + \tau^3{}_{2,2} + \tau^3{}_{3,3} = \tau^3{}_{j,j} = -\,F^3 \qquad (15.14c)$$

Or simply, $\quad \tau^i{}_{1,1} + \tau^i{}_{2,2} + \tau^i{}_{3,3} = \tau^i{}_{j,j} = -\,F^i \qquad (15.14)$

Here, the body force component $-F^i$ is associated with the inertial force component $\mu_0.\partial^2 u^i/\partial t^2$, where μ_0 is inertial constant for the Elastic Continuum and $\partial^2 u^i/\partial t^2$ represents the acceleration corresponding to u^i. Therefore, the equilibrium equation (15.14) may be written as,

$$\tau^i{}_{1,1} + \tau^i{}_{2,2} + \tau^i{}_{3,3} = \tau^i{}_{j,j} = \mu_0.\partial^2 u^i/\partial t^2 \qquad (15.15A)$$

In orthogonal curvilinear coordinates with metric tensor components g^{ij}, the general equilibrium equations for the Elastic Continuum take the form,

$$g^{11}\tau^i{}_{1,1} + g^{22}\tau^i{}_{2,2} + g^{33}\tau^i{}_{3,3} = g^{jj}\tau^i{}_{j,j} = \mu_0.\partial^2 u^i/\partial t^2 \qquad (15.15)$$

15.3 Stress Strain Relations in the Elastic Space Continuum

Modified Hooke's Law. In the generalized Hooke's law for elastic material bodies, the effect of 'atomicity' or structural discreteness gets accommodated through the Poisson's ratio constant. Further, the effect of a finite value of Poisson's ratio constant for a material body is manifested through different values of speed of propagation of transverse and longitudinal strain waves. Therefore, in contrast to an elastic material body, we shall take the Poisson's ratio constant for the Elastic Space Continuum to be zero to ensure same speed of propagation of transverse and longitudinal strain waves. With this, the generalized Hooke's law will get modified to a simple form as,

$$\tau^i_j = (1/\varepsilon_0).S^i_j = (1/\varepsilon_0).u^i_{,j} \qquad (15.16)$$

where $(1/\varepsilon_0)$ is the elastic constant for the Elastic Space Continuum, in appropriate units. In conventional electrical units, dimensions of $(1/\varepsilon_0)$ are $Nm^2/Coul^2$. However, in mechanical units, the dimensions of elastic constant $(1/\varepsilon_0)$ are required to be N/m^2. Therefore, to ensure the compatibility of electrical and mechanical units in the Elastic Continuum, we must assign the dimension of $[M^0L^2T^0]$ or m^2 to the electrical unit of charge - Coulomb. A rough estimate for the equivalence of Coulomb is that one Coulomb is equal to the square of Compton wavelength of the electron. That is, one Coulomb is equal to 5.887×10^{-24} m^2. The corresponding value of the elastic constant $(1/\varepsilon_0)$ works out to be about 3.259×10^{57} N/m^2.

Substituting this relation (15.16) in the dynamic equilibrium equation (15.15), we get the corresponding equilibrium equation in terms of displacement components u^i as,

$$(1/\varepsilon_0).[\, g^{11}u^i_{,11} + g^{22}u^i_{,22} + g^{33}u^i_{,33}\,] = (1/\varepsilon_0).\, g^{jj}u^i_{,jj} = \mu_0.\partial^2u^i/\partial t^2 \qquad (15.17A)$$

Or

$$g^{11}u^i_{,11} + g^{22}u^i_{,22} + g^{33}u^i_{,33} = g^{jj}u^i_{,33} = \varepsilon_0\mu_0.\partial^2u^i/\partial t^2 = (1/c^2)\, \partial^2u^i/\partial t^2 \qquad (15.17)$$

Thus the dynamic equilibrium equation for the Elastic Space Continuum comes out to be the standard vector wave equation involving displacement vector components u^1, u^2 and u^3.

In conventional Cartesian coordinate system (x,y,z), with physical components of the displacement vector **U** given by u^x, u^y and u^z equation (15.17) reduces to a set of three second order partial differential equations as:

$$\partial^2 u^x/\partial x^2 + \partial^2 u^x/\partial y^2 + \partial^2 u^x/\partial z^2 = (1/c^2)\, \partial^2 u^x/\partial t^2$$
$$\partial^2 u^y/\partial x^2 + \partial^2 u^y/\partial y^2 + \partial^2 u^y/\partial z^2 = (1/c^2)\, \partial^2 u^y/\partial t^2$$
$$\partial^2 u^z/\partial x^2 + \partial^2 u^z/\partial y^2 + \partial^2 u^z/\partial z^2 = (1/c^2)\, \partial^2 u^z/\partial t^2$$

These three equations may be grouped into one equation involving vector **U** as,

$$\partial^2 \mathbf{U}/\partial x^2 + \partial^2 \mathbf{U}/\partial y^2 + \partial^2 \mathbf{U}/\partial z^2 = \nabla^2 \mathbf{U} = (1/c^2)\, \partial^2 \mathbf{U}/\partial t^2 \qquad (15.18)$$

Equations (15.18) are the standard equilibrium equations of elasticity in the Elastic Space Continuum which can be written in any convenient coordinate system.

15.4 Equilibrium Equations in Different Coordinate Systems

The general equilibrium equation (15.17) given in tensor notation can be easily adapted to any particular coordinate system with metric tensor g^{ij}. Rewriting this equation, we have

$$g^{11}u^i_{,11} + g^{22}u^i_{,22} + g^{33}u^i_{,33} = g^{jj}u^i_{,jj} = (1/c^2)\, \partial^2 u^i/\partial t^2 \qquad (15.19)$$

where the terms $u^i_{,jj}$ represent the second order covariant derivative of displacement vector u^i. However, for physical applications we have to finally convert all covariant and contravariant tensor components to their corresponding physical components. Some of the important steps that are relevant for adaptation of the tensor equations of elasticity to spherical polar, cylindrical or any other orthogonal coordinate system, involving physical components of displacement vector **U**, are given below.

(a) The covariant derivative of u^i is given by,

$$u^i_{,j} = \partial u^i/\partial x^j + \Gamma^i_{\alpha j}\, u^\alpha \qquad \text{(summation over } \alpha) \qquad (15.20)$$

where, Γ^i_{jk} are the Christoffel symbols of second kind.

(b) The second covariant derivative of $u^i_{,j}$ is given by

$$u^i_{,jj} = \frac{\partial u^i_{,j}}{\partial x^j} + \Gamma^i_{\alpha j}\, u^\alpha_{,j} - \Gamma^\alpha_{jj}\, u^i_{,\alpha} \qquad \text{(summation over } \alpha) \qquad (15.21)$$

(c) Physical components of strain, which must be dimensionless, are given by

$$S^{x^i}_{x^j} = \sqrt{g_{ii}}\,.\,u^i_{,j}\,.\,\sqrt{g^{jj}} \qquad (15.22)$$

(d) The physical components of displacement vector **U**, which must have the dimensions of length [L], corresponding to the contravariant components u^i are given by

$$u^{x^i} = \sqrt{g_{ii}}\,.\,u^i \qquad (15.23)$$

(e) The physical components of temporal strain, which again must be dimensionless, corresponding to the time derivative of contravariant components u^i, are given by

$$S_t^{x^i} = \frac{1}{c}\frac{\partial u^{x^i}}{\partial t} = \sqrt{g_{ii}}\,\frac{1}{c}\frac{\partial u^i}{\partial t} \tag{15.24}$$

Spherical Polar Coordinates. Let us now consider a spherical polar coordinate system given by $x^1 = r$, $x^2 = \theta$ and $x^3 = \phi$ coordinates, related to conventional Cartesian coordinates x, y, z as

$$x = r\sin\theta\cos\phi \quad ; \quad y = r\sin\theta\sin\phi \quad ; \quad z = r\cos\theta \tag{15.25}$$

The non-zero metric tensor components g_{ij} and g^{ij} for this coordinate system are,

$$g_{11} = 1 \;;\quad g_{22} = r^2 \;;\quad g_{33} = r^2\sin^2\theta \tag{15.26A}$$

and

$$g^{11} = 1 \;;\quad g^{22} = 1/r^2 \;;\quad g^{33} = 1/(r^2\sin^2\theta) \tag{15.26B}$$

The corresponding Christoffel symbols of second kind $\Gamma^i{}_{jk}$ are given by,

$$\Gamma^1{}_{22} = -r \quad ; \quad \Gamma^2{}_{12} = \Gamma^2{}_{21} = 1/r \quad ; \quad \Gamma^1{}_{33} = -r\sin^2\theta$$

and

$$\Gamma^2{}_{33} = -\sin\theta\cos\theta \;;\; \Gamma^3{}_{13} = \Gamma^3{}_{31} = 1/r \;;\; \Gamma^3{}_{23} = \Gamma^3{}_{32} = \cot\theta \tag{15.27}$$

The physical components u^r, u^θ, u^ϕ of displacement vector **U** are related to the corresponding contravariant components u^1, u^2, u^3 through equation (15.23) as,

$$u^r = u^1 \;;\quad u^\theta = r\,u^2 \;;\quad u^\phi = r\sin\theta\, u^3 \tag{15.28}$$

The physical components of spatial strain are obtained from equation (15.20) & (15.22) as:

$$S_r^r = \frac{\partial u^r}{\partial r} \;;\quad S_\theta^r = \frac{1}{r}\cdot\frac{\partial u^r}{\partial \theta} - \frac{u^\theta}{r} \;;\quad S_\phi^r = \frac{1}{r.\sin\theta}\cdot\frac{\partial u^r}{\partial \phi} - \frac{u^\phi}{r}$$

$$S_r^\theta = \frac{\partial u^\theta}{\partial r} \;;\quad S_\theta^\theta = \frac{1}{r}\cdot\frac{\partial u^\theta}{\partial \theta} + \frac{u^r}{r} \;;\quad S_\phi^\theta = \frac{1}{r.\sin\theta}\cdot\frac{\partial u^\theta}{\partial \phi} - \frac{\cot\theta}{r}.u^\phi$$

$$S_r^\phi = \frac{\partial u^\phi}{\partial r} \;;\quad S_\theta^\phi = \frac{1}{r}\cdot\frac{\partial u^\phi}{\partial \theta} \;;\quad S_\phi^\phi = \frac{1}{r.\sin\theta}\cdot\frac{\partial u^\phi}{\partial \phi} + \frac{\cot\theta}{r}.u^\theta + \frac{u^r}{r}$$

And the corresponding physical components of temporal strain are given by:

$$S_t^r = \frac{1}{c}\cdot\frac{\partial u^r}{\partial t} \;;\quad S_t^\theta = \frac{1}{c}\cdot\frac{\partial u^\theta}{\partial t} \;;\quad S_t^\phi = \frac{1}{c}\cdot\frac{\partial u^\phi}{\partial t} \tag{15.29}$$

The dynamic equilibrium equations (15.17) given in tensor notation can now be rewritten in terms of physical components (u^r, u^θ, u^ϕ) of displacement vector **U**, in spherical polar coordinates, by using equations (15.20) to (15.29) above, as follows

$$\frac{\partial^2 u^r}{\partial r^2}+\frac{2}{r}.\frac{\partial u^r}{\partial r}-\frac{2}{r^2}.u^r+\frac{1}{r^2}\left(\frac{\partial^2 u^r}{\partial \theta^2}+\cot\theta.\frac{\partial u^r}{\partial \theta}+\frac{1}{\sin^2\theta}.\frac{\partial^2 u^r}{\partial \phi^2}\right)$$
$$-\frac{2}{r^2}\left(\frac{\partial u^\theta}{\partial \theta}+\cot\theta.u^\theta+\frac{1}{\sin\theta}.\frac{\partial u^\phi}{\partial \phi}\right)=\frac{1}{c^2}.\frac{\partial^2 u^r}{\partial t^2}$$

$$\frac{\partial^2 u^\theta}{\partial r^2}+\frac{2}{r}.\frac{\partial u^\theta}{\partial r}-\frac{u^\theta}{r^2.\sin^2\theta}+\frac{1}{r^2}\left(\frac{\partial^2 u^\theta}{\partial \theta^2}+\cot\theta.\frac{\partial u^\theta}{\partial \theta}+\frac{1}{\sin^2\theta}.\frac{\partial^2 u^\theta}{\partial \phi^2}\right)$$
$$+\frac{2}{r^2}\left(\frac{\partial u^r}{\partial \theta}-\frac{\cot\theta}{\sin\theta}.\frac{\partial u^\phi}{\partial \phi}\right)=\frac{1}{c^2}.\frac{\partial^2 u^\theta}{\partial t^2}$$

$$\frac{\partial^2 u^\phi}{\partial r^2}+\frac{2}{r}.\frac{\partial u^\phi}{\partial r}-\frac{u^\phi}{r^2.\sin^2\theta}+\frac{1}{r^2}\left(\frac{\partial^2 u^\phi}{\partial \theta^2}+\cot\theta.\frac{\partial u^\phi}{\partial \theta}+\frac{1}{\sin^2\theta}.\frac{\partial^2 u^\phi}{\partial \phi^2}\right)$$
$$+\frac{2}{r^2.\sin\theta}\left(\frac{\partial u^r}{\partial \phi}+\cot\theta.\frac{\partial u^\theta}{\partial \phi}\right)=\frac{1}{c^2}.\frac{\partial^2 u^\phi}{\partial t^2}$$

... (15.30)

The equilibrium equations (15.30) constitute a set of three simultaneous partial differential equations involving displacement vector components u^r, u^θ and u^ϕ. Unlike the case of equilibrium equations (15.18) in conventional Cartesian coordinate system, these equations in spherical polar coordinates may be considered 'mutually coupled' in the sense that none of these equations can be solved independent of one another.

Cylindrical Coordinates. In a cylindrical coordinate system, defined by $x^1 = \rho$, $x^2 = \phi$ and $x^3 = z$, related to conventional Cartesian coordinates x, y, z as,

$$x = \rho\cos\phi \quad ; \quad y = \rho\sin\phi \quad ; \quad z = z \tag{15.31}$$

The non-zero metric tensor components g_{ij} and g^{ij} for this coordinate system are:

$$g_{11} = 1 \quad ; \quad g_{22} = \rho^2 \quad ; \quad g_{33} = 1 \tag{15.32A}$$

and $$g^{11} = 1 \quad ; \quad g^{22} = 1/\rho^2 \quad ; \quad g^{33} = 1 \tag{15.32B}$$

The corresponding Christoffel symbols of second kind Γ^i_{jk} are given by,

$$\Gamma^1_{22} = -\rho \quad ; \quad \Gamma^2_{12} = \Gamma^2_{21} = 1/\rho \tag{15.33}$$

The physical components u^ρ, u^ϕ, u^z of displacement vector **U** are related to the corresponding contravariant components u^1, u^2, u^3 through equation (15.23) as,

$$u^\rho = u^1 \quad ; \quad u^\phi = \rho\, u^2 \quad ; \quad u^z = u^3 \tag{15.34}$$

The physical components of spatial strain are obtained from equations (15.20) & (15.22) as:

$$S^\rho_\rho = \frac{\partial u^\rho}{\partial \rho} \quad ; \quad S^\rho_\phi = \frac{1}{\rho}\cdot\frac{\partial u^\rho}{\partial \phi} - \frac{u^\phi}{\rho} \quad ; \quad S^\rho_z = \frac{\partial u^\rho}{\partial z}$$

$$S^\phi_\rho = \frac{\partial u^\phi}{\partial \rho} \quad ; \quad S^\phi_\phi = \frac{1}{\rho}\cdot\frac{\partial u^\phi}{\partial \phi} + \frac{u^\rho}{\rho} \quad ; \quad S^\phi_z = \frac{\partial u^\phi}{\partial z}$$

$$S^z_\rho = \frac{\partial u^z}{\partial \rho} \quad ; \quad S^z_\phi = \frac{1}{\rho}\cdot\frac{\partial u^z}{\partial \phi} \quad ; \quad S^z_z = \frac{\partial u^z}{\partial z}$$

And the corresponding physical components of temporal strain are given by:

$$S^\rho_t = \frac{1}{c}\cdot\frac{\partial u^\rho}{\partial t} \quad ; \quad S^\phi_t = \frac{1}{c}\cdot\frac{\partial u^\phi}{\partial t} \quad ; \quad S^z_t = \frac{1}{c}\cdot\frac{\partial u^z}{\partial t} \tag{15.35}$$

The dynamic equilibrium equations (15.17) can now be rewritten in terms of physical components (u^ρ, u^ϕ, u^z) of displacement vector **U**, in cylindrical coordinates as follows.

$$\frac{\partial^2 u^\rho}{\partial \rho^2} + \frac{1}{\rho}\cdot\frac{\partial u^\rho}{\partial \rho} - \frac{u^\rho}{\rho^2} + \frac{1}{\rho^2}\cdot\frac{\partial^2 u^\rho}{\partial \phi^2} + \frac{\partial^2 u^\rho}{\partial z^2} - \frac{2}{\rho^2}\cdot\frac{\partial u^\phi}{\partial \phi} = \frac{1}{c^2}\cdot\frac{\partial^2 u^\rho}{\partial t^2}$$

$$\frac{\partial^2 u^\phi}{\partial \rho^2} + \frac{1}{\rho}\cdot\frac{\partial u^\phi}{\partial \rho} - \frac{u^\phi}{\rho^2} + \frac{1}{\rho^2}\cdot\frac{\partial^2 u^\phi}{\partial \phi^2} + \frac{\partial^2 u^\phi}{\partial z^2} + \frac{2}{\rho^2}\cdot\frac{\partial u^\rho}{\partial \phi} = \frac{1}{c^2}\cdot\frac{\partial^2 u^\phi}{\partial t^2}$$

$$\frac{\partial^2 u^z}{\partial \rho^2} + \frac{1}{\rho}\cdot\frac{\partial u^z}{\partial \rho} + \frac{1}{\rho^2}\cdot\frac{\partial^2 u^z}{\partial \phi^2} + \frac{\partial^2 u^z}{\partial z^2} = \frac{1}{c^2}\cdot\frac{\partial^2 u^z}{\partial t^2} \tag{15.36}$$

Equilibrium equations (15.36) constitute a set of three simultaneous partial differential equations involving displacement vector components u^ρ, u^ϕ, and u^z.

16

Strain Waves and the Electro Magnetic Field

16.1 Strain Wave Propagation in Elastic Space Continuum

In conventional Cartesian coordinate system (x,y,z) with physical components of the displacement vector **U** given by u^x, u^y and u^z, the equilibrium equations of elasticity are given by a set of three second order partial differential equations as:

$$\partial^2 u^x/\partial x^2 + \partial^2 u^x/\partial y^2 + \partial^2 u^x/\partial z^2 = (1/c^2)\ \partial^2 u^x/\partial t^2 \qquad (16.1A)$$

$$\partial^2 u^y/\partial x^2 + \partial^2 u^y/\partial y^2 + \partial^2 u^y/\partial z^2 = (1/c^2)\ \partial^2 u^y/\partial t^2 \qquad (16.1B)$$

$$\partial^2 u^z/\partial x^2 + \partial^2 u^z/\partial y^2 + \partial^2 u^z/\partial z^2 = (1/c^2)\ \partial^2 u^z/\partial t^2 \qquad (16.1C)$$

These three equations may be grouped into one equation involving vector **U** as,

$$\partial^2\mathbf{U}/\partial x^2 + \partial^2\mathbf{U}/\partial y^2 + \partial^2\mathbf{U}/\partial z^2 = \nabla^2\mathbf{U} = (1/c^2)\ \partial^2\mathbf{U}/\partial t^2 \qquad (16.1)$$

In the above equation, the displacement vector field **U** may be expressed as a combination of two functions - a vector function **A**(x,y,z,t) and a scalar function ϕ(x,y,z,t) as,

$$\mathbf{U} = \nabla\times\mathbf{A} + \nabla\phi \qquad (16.2)$$

Here, if the scalar function $\phi = 0$, then

$$\nabla.\mathbf{U} = 0 \qquad (16.3)$$

And the vector function **A** will then satisfy equation (16.1) as,

$$\partial^2\mathbf{A}/\partial x^2 + \partial^2\mathbf{A}/\partial y^2 + \partial^2\mathbf{A}/\partial z^2 = \nabla^2\mathbf{A} = (1/c^2)\ \partial^2\mathbf{A}/\partial t^2 \qquad (16.4)$$

The equations (16.1), (16.3) and (16.4) represent a solenoidal or transverse strain wave propagation through the Elastic Space Continuum and the displacement vector field **U** may be termed as a solenoidal vector field. If on the other hand vector function **A**= 0, then $\nabla\times\mathbf{U} = 0$ and the scalar function ϕ will satisfy equation (16.1) as,

$$\partial^2\phi/\partial x^2 + \partial^2\phi/\partial y^2 + \partial^2\phi/\partial z^2 = \nabla^2\phi = (1/c^2)\ \partial^2\phi/\partial t^2 \qquad (16.5)$$

In this case, equations (16.1), (16.2) & (16.5) will represent irrotational or longitudinal strain wave propagation through the Space. In both of these cases, the spatial strain components as functions of space-time coordinates will be given by the terms $\partial u^x/\partial x$, $\partial u^x/\partial y$, $\partial u^x/\partial z$, $\partial u^y/\partial x$, $\partial u^y/\partial y$, $\partial u^y/\partial z$, $\partial u^z/\partial x$, $\partial u^z/\partial y$ and $\partial u^z/\partial z$, whereas the temporal strain components as functions of space and time coordinates will be given by the terms $(1/c)\partial u^x/\partial t$, $(1/c)\partial u^y/\partial t$ and $(1/c)\partial u^z/\partial t$.

16.2 Inertial Property of the Space Continuum

The displacement vector components u^i, in general will be functions of time and space coordinates. Whereas the covariant derivatives $u^i_{,j}$ represent spatial strain components, the partial derivatives of u^i with respect to time t, that is, $(1/c).\partial u^i/\partial t$ constitute temporal strain components. If we regard time as a fourth dimension coordinate in quadrature to the three space coordinates, we may study the dynamic phenomenon occurring in 3-D space through the geometry of 4-D space-time manifold. Viewing the above mentioned spatial and temporal strain components as occurring in the four dimensional space-time construct, we see that,

$$S^i_4 = u^i_{,4} = \partial u^i/\partial x^4 = (1/i\, c).\partial u^i/\partial t \qquad \text{where } x^4 = i\, ct$$

and $$S^i_j = u^i_{,j} \qquad (i \rightarrow 1 \text{ to } 3 \;\; \& \;\; j \rightarrow 1 \text{ to } 4)$$

With conventional Cartesian coordinate system (x,y,z), let the fourth coordinate x^4 be represented by η such that $\partial u^i/\partial \eta = (1/ic).\partial u^i/\partial t$. The inertial term in equation (1) will change to $(1/c^2)\partial^2\mathbf{U}/\partial t^2 = -\,\partial^2\mathbf{U}/\partial \eta^2$. Accordingly the dynamic equilibrium equation (1) will transform to,

$$\partial^2\mathbf{U}/\partial x^2 + \partial^2\mathbf{U}/\partial y^2 + \partial^2\mathbf{U}/\partial z^2 + \partial^2\mathbf{U}/\partial \eta^2 = \square^2\mathbf{U} = 0 \qquad (16.6)$$

This shows that in the four dimensional representation of dynamic equilibrium equations, the inertial term is no longer explicit. The inertial constant μ_0 may be depicted in terms of the elastic constant $1/\varepsilon_0$ and c as $\mu_0 = (1/\varepsilon_0).(1/c^2)$. As such, a finite (non-zero) value of the inertial constant μ_0 may be attributed to the finite (less than infinite) value of c in the space-time manifold. Hence, the inertial property of the Elastic Space Continuum may be viewed as a consequence of finite velocity of light. In fact, even the dynamic equilibrium equation (16.6), in the four dimensional space-time may be derived ab-initio without invoking the concept of inertial body force. This indicates that the dynamic deformations and interactions occurring in the elastic space continuum can be studied as a geometrical phenomenon in the abstract space-time manifold without invoking the notion of inertial body forces in the continuum. However, on logical grounds, we will prefer to study physical reality as *space dynamics* rather than as *geometry of space-time*.

16.3 Electromagnetic Field in the Elastic Space Continuum

Notion of Field in Physics. A field is defined as an assignment of a physical quantity to every point in space. A field thus extends throughout a large region of space so that its influence is all-pervading.

Usually the strength of a field varies over a region. Fields are usually represented mathematically by scalar, vector and tensor fields. Michael Faraday was the first to realize that electric and magnetic fields are not only fields of force which dictate the motion of particles, but also have an independent physical reality because they carry energy. These ideas eventually led to the creation of a unified field theory in physics with the development of Maxwell's electromagnetic field equations. However, the current representation of ***field*** in physics does not provide sufficient insight regarding the *real physical entity* that is supposed to *exist* at every point in space.

The electromagnetic (EM) field is a physical field produced by electrically charged objects. It affects the behavior of charged objects in its vicinity. The electromagnetic field extends indefinitely throughout space and describes the electromagnetic interaction. An electromagnetic field can be viewed as a combination of an electric field **E**(x,y,z,t) and a magnetic field **B**(x,y,z,t). The electric field is produced by stationary charges, and the magnetic field by currents (moving charges), which are often described as the sources of the field.

The Maxwell's equations for the electromagnetic field in free space, take the following form in an area that is very far away from any charges or currents.

$$\nabla.\mathbf{E} = 0 \qquad (16.7)$$

$$\nabla.\mathbf{B} = 0 \qquad (16.8)$$

$$\nabla\times\mathbf{E} = -\ \partial\mathbf{B}/\partial t \qquad (16.9)$$

$$\nabla\times\mathbf{B} = (1/c^2).\ \partial\mathbf{E}/\partial t \qquad (16.10)$$

The transverse electromagnetic waves in free space, characterized by zero divergence are represented by the following standard wave equations,

$$\nabla^2\mathbf{E} = (1/c^2)\ \partial^2\mathbf{E}/\partial t^2 \qquad (16.11)$$

$$\nabla^2\mathbf{B} = (1/c^2)\ \partial^2\mathbf{B}/\partial t^2 \qquad (16.12)$$

Here, the equations (16.7), (16.8), (16.11) and (16.12) above are identical in form to the equations (16.1), (16.3) and (16.4), representing solenoidal or transverse strain wave propagation through an Elastic Continuum. This identity in '*form*' may be extended to an identity in '*essence*' through the following correlation between displacement vector field **U** (or the corresponding temporal and spatial strain components) and the electromagnetic field vectors **E** and **B** ,

$$\mathbf{E} = -\ (1/\varepsilon_0).(1/c).\partial\mathbf{U}/\partial t \qquad (16.13)$$

$$\mathbf{B} = (1/c).(1/\varepsilon_0).\ (\nabla\times\mathbf{U}) \qquad (16.14)$$

Through this correlation, in conjunction with equation (16.3), the electromagnetic field equations (16.7) to (16.12) are also satisfied identically. In essence, this means that the electric field vector **E** represents the 'temporal stress' field in the Elastic Space Continuum and is always a function of space and time coordinates. On the other hand, the magnetic field vector **B** represents (1/c) times the 'torsional stress' in the Elastic Space Continuum and is also a function of space and time coordinates. Therefore, as a logical consequence of comparing the propagation of transverse strain waves and the electromagnetic waves, in the elastic space continuum, the electromagnetic field in the so called 'vacuum' or empty space, is found to be a dynamic stress-strain field in the corresponding Elastic Space Continuum.

From equation (16.13) above, it can also be seen that Maxwell's electric displacement **D,** given by $\mathbf{D} = -(1/c).\partial\mathbf{U}/\partial t$, actually represents temporal strain component in the Elastic Continuum. One most pertinent point to be noted here, is that at any given point in the continuum, the displacement vector **U** and the strain tensor **S** provide more complete information regarding the physical state of the continuum than do the electromagnetic field vectors **E** and **B**.

The above mentioned stress-strain tensor concepts are mainly associated with electromagnetic field vectors defined in matter free space. The unit volt/m identified with electric field vector **E** is seen to be equivalent to Joule/Coulomb.m or Newton/Coulomb which as per the earlier remarks on dimensions, can be further reduced to N/m^2 - a unit of physical stress in the elastic continuum. However, in a region of space influenced by the presence of electric charges in the vicinity, one component of electric field vector **E** is obtained as a gradient of Coulomb potential ϕ, which is essentially an interaction parameter. The Coulomb interaction potential ϕ, as will be seen later, is a consequence of or the end result of mutual interactions among various charged particles. Thus, the electric field **E** obtained as a gradient of ϕ, represents an interaction force acting on mutually interacting charged particles and is strictly not the same thing as physical stress in the elastic continuum. But the equivalence of the practical units of **E** representing the physical dynamic stress and those of **E** representing mutual interaction force among charged particles, permits us to use both these concepts side by side without much distinction.

16.4 Energy Density in the Strain Fields

In the deformed or stressed state of the Elastic Space Continuum a certain amount of strain energy will get stored in the region under stress.

The strain energy density W at a point P of the continuum, will obviously be a function of the intensity of strain at that point. Since the strain energy stored in any arbitrarily small volume δV of the Continuum under stress, has to be positive, the strain energy density function W will be a positive definite form of the strain components S^i_j. Further, this strain energy density W or the energy of deformation per unit volume, has a physical meaning that is independent of the choice of coordinate system and hence is an invariant. Therefore, using the Clapeyron formula for the strain energy density for ordinary material bodies under static equilibrium, expressed in conventional Cartesian coordinate system, the spatial strain energy density for the Elastic Space Continuum may be given by,

$$\begin{aligned} W_s &= \tfrac{1}{2}\,\tau^i_j\, S^i_j = \tfrac{1}{2}\,(1/\varepsilon_0)\, S^i_j\, S^i_j \quad \text{(summation over i, j} \rightarrow \text{1 to 3)} \\ &= \tfrac{1}{2}\,(1/\varepsilon_0)\,[\,(S^1{}_1)^2 + (S^2{}_2)^2 + (S^3{}_3)^2 + (S^1{}_2)^2 \\ &\quad + (S^2{}_3)^2 + (S^3{}_2)^2 + (S^2{}_1)^2 + (S^1{}_3)^2 + (S^3{}_1)^2] \end{aligned} \tag{16.15}$$

This formula for the strain energy density function W will also hold good in all other orthogonal curvilinear coordinate systems, provided we use physical strain components in place of tensor components S^i_j. In a material body if the strain intensity varies with time, the kinetic energy density is given by $\tfrac{1}{2}\,\rho(\partial u^i/\partial t).(\,\partial u^i/\partial t)$. Therefore, the temporal strain energy density in the Elastic Space Continuum will be given by,

$$\begin{aligned} W_t &= \frac{\mu_0}{2}.\frac{\partial u^i}{\partial t}.\frac{\partial u^i}{\partial t} = \frac{1}{2\varepsilon_0}.\left(\frac{1}{c}\frac{\partial u^i}{\partial t}\right).\left(\frac{1}{c}\frac{\partial u^i}{\partial t}\right) \quad \text{(summation over i} \rightarrow \text{1 to 3)} \\ &= (1/2\varepsilon_0).S^i_t.\,S^i_t \end{aligned} \tag{16.16}$$

Hence, the total strain energy density W within a particular strain field region of the Continuum will be given by,

$$\begin{aligned} W &= W_s + W_t \\ &= (1/2\varepsilon_0).[S^i_j\, S^i_j + S^i_t.\, S^i_t\,] \quad \text{(summation over i, j} \rightarrow \text{1 to 3)} \end{aligned} \tag{16.17}$$

However, the above equation (16.17) for the strain energy density is strictly valid only when the temporal strain components are in quadrature to the corresponding spatial strain components. When displacement components u^i involve space and time coordinates as independent parameters, representing standing wave oscillations, the temporal strain components will be in quadrature to the corresponding spatial strain components. On the other hand, when u^i involve functionally interlinked space and time coordinates, representing propagating phase waves, the temporal strain components may then assume phase opposition to the corresponding spatial strain components.

Let us consider a particular solution of displacement components u^i that involve a propagating phase wave function of the type $e^{i(\kappa x^j \pm \kappa ct)}$ then the corresponding spatial strain terms $\partial u^i/\partial x^j$ or $u^i_{,j}$ will be in phase opposition to the temporal strain terms $(1/c).\partial u^i/\partial t$. In such cases we may introduce a space-time phase parameter ψ given by,

$$\psi = \kappa x^j + \kappa ct \tag{16.18}$$

such that, $$d\psi = \frac{\partial \psi}{\partial x^j}.dx^j + \frac{\partial \psi}{\partial t}.dt \quad \text{(no summation)} \tag{16.19}$$

For surfaces of constant phase in the strain field, representing phase wave propagation $d\psi = 0$ and from equation (16.19),

$$\frac{\partial \psi}{\partial x^j}.dx^j = -\frac{\partial \psi}{\partial t}.dt \tag{16.20}$$

The $f(x).e^{i\psi}$ terms in u^i, where $f(x)$ is any function of space coordinates alone, will also represent the surfaces of constant phase propagating along x^j coordinate. The effective total strain component S^i_j for such a case of propagating phase waves, where x^j and t are interlinked through ψ, will be given by,

$$S^i_j = \frac{du^i}{dx^j} = \frac{\partial u^i}{\partial x^j} + \frac{\partial u^i}{\partial t}.\frac{dt}{dx^j} = \frac{\partial u^i}{\partial x^j} + \frac{\partial u^i}{\partial t}.\left(-\frac{\partial \psi/\partial x^j}{\partial \psi/\partial t}\right) \quad \text{(by using (16.20))}$$

$$= \frac{\partial u^i}{\partial x^j} - \frac{1}{c}.\frac{\partial u^i}{\partial t} = f'(x).e^{i\psi} + f(x).[\kappa - \kappa c/c].\ e^{i\psi}$$

$$= f'(x).e^{i\psi} + 0 \tag{16.21}$$

That is, the temporal strain component gets subtracted from the corresponding spatial strain component. In other words, for the strain field consisting of phase waves of the type $u^i = f(x).e^{i\psi}$ propagating along x^j coordinate, the effective temporal strain for displacement component u^i will be in phase opposition to the corresponding spatial strain component S^i_j. In effect, this implies that the $e^{i\psi}$ type terms occurring in u^i will not contribute anything in the effective total strain. Hence, for computing the total strain energy density in such cases, the $e^{i\psi}$ type terms occurring in various displacement components, may be treated as constants. The total strain energy density in phase wave fields discussed above, will therefore depend only on the amplitude term $f'(x)$ or more precisely, on the rms (root mean square) value of the amplitude term of such strain waves. We shall encounter such phase wave fields in the study of electrostatic field of charged particles.

17

Solutions of Equilibrium Equations

17.1 Constraints on Solution of Equilibrium Equations

When any region of the Elastic Space Continuum is subjected to some sort of deformation, a strain field may be said to have developed in that region. This strain field is fully defined, when the displacement vector **U** is completely determined as a function of space and time coordinates over the whole region of the Continuum under deformation. The displacement vector components u^i can be completely determined from the detailed solution of the equilibrium equations,

$$g^{jj}u^i_{,jj} = (1/c^2)\,\partial^2 u^i/\partial t^2 \qquad (17.1A)$$

$$\nabla^2 \mathbf{U} = (1/c^2)\,\partial^2 \mathbf{U}/\partial t^2 \qquad (17.1B)$$

Hence, the detailed study of any deformed or stressed region of the Elastic Space Continuum primarily involves the detailed solution of the equilibrium equations subject to appropriate boundary conditions. Unlike ordinary linear differential equations, the general solutions of partial differential equations contain arbitrary functions which are difficult to adjust, so as to satisfy the given boundary conditions. For different sets of boundary conditions, the given partial differential equations will yield different unique solutions.

However, most of the boundary value problems involving linear partial differential equations, can be solved by the method of separation of variables. It involves a solution in a particular coordinate system, which breaks up into a product of functions, each of which contains only one of the independent coordinate parameters. In a particular coordinate system, if the boundary conditions characterizing a given physical situation are such that the corresponding unique solution for u^i consists of a product of functions, each of which contains only one of the independent variables, then the boundary conditions may be said to be 'symmetric' in that coordinate system. The method of separation of variables is applicable for the solution of equilibrium equations in a given coordinate system, if the boundary conditions are 'symmetric' in that coordinate system. Therefore, depending on 'symmetry' of the boundary conditions, an appropriate coordinate system will be used for solution of the equilibrium equations.

General Boundary Conditions. Let V be the total volume and Σ be the outer boundary surface of a particular region of the Elastic Space Continuum under deformation. The general boundary conditions that must be satisfied by the displacement vector components u^i obtained from the solution of equilibrium equations (17.1), may be listed as:

- The displacement components u^i must vanish at the boundary Σ and must remain finite and continuous within this boundary. The symmetry of boundary conditions in a particular coordinate system will be governed by the shape of Σ.
- The strain components and the strain energy density must be finite and continuous within the boundary Σ of the region under consideration. On the boundary Σ the stress and hence the strain components may either vanish or be finite, periodic and preferably symmetric with respect to the center of the region, such that at any instant the surface integral of the stress vector over Σ must vanish.
- The total strain energy within the entire volume V must be finite and remain constant or time invariant in the absence of any external interaction.
- The amplitude of displacement vector components u^i will be proportional to the wave angular frequency $\omega=2\pi\nu$ or its equivalent parameter $\omega/c=2\pi/\lambda=\kappa$ which is the wave number of the strain wave oscillations occurring within the entire volume V of the Continuum under stress. This is due to the fact that whenever the amplitude of displacement vector **U** starts building up in any region of the Continuum, it will simultaneously start dissipating or spreading out to its surroundings at velocity c. Therefore, a higher magnitude of displacement vector amplitude will result whenever the rate of build up of **U** is high in comparison to c. However, this condition may be taken as a postulate at this stage.
- Since the dimension of displacement vector **U** has to be $[M^0L^1T^0]$, we will take the integration constant for u^i as a dimensionless constant multiplied by $e\kappa$, where e is the magnitude of electron charge in Coulombs. With this we will use the elastic constant $(1/\varepsilon_0)$ in the units of $Nm^2/Coul^2$.

17.2 Strain Bubbles in the Elastic Space Continuum

Types and distinguishing features of Strain Bubbles. A closed region of the Elastic Space Continuum, that contains a finite amount of energy stored in its strain field satisfying appropriate boundary conditions, may be called a 'Strain Bubble'. From the nature of general boundary conditions and the equilibrium equations, it turns out that all valid solutions for displacement vector components u^i are functions of space and time coordinates representing various types of strain wave oscillations. That is, all 'Strain Bubbles' contain a constant, finite amount of total strain energy and essentially consist of various strain wave oscillations within a specific boundary surface Σ of the Elastic Space Continuum. Three main distinguishing features of various types of strain bubbles are,

- ***Shape and symmetry of boundary surface Σ.*** The shape of the boundary surface Σ where the components u^i vanish altogether, governs the shape and to some extent the size of the strain bubble. If Σ is the surface of a right circular cylinder, the corresponding strain bubble may be called a 'Cylindrical Strain Bubble'. If Σ is a spherical surface, the strain bubble may be termed as a 'Spherical Strain Bubble', and corresponding to rectangular box shape of Σ, the strain bubble may be referred to as a 'Cartesian Strain Bubble'. Therefore, from the foregoing discussions about the symmetry of Σ, it is obvious that cylindrical strain bubble solutions will be obtained from the equilibrium equations (17.1) written in cylindrical coordinates. Similarly, spherical and Cartesian strain bubble solutions will be obtained from equilibrium equations written in spherical and Cartesian coordinate systems respectively.
- ***Size of the Boundary Surface Σ.*** If the boundary surface Σ is located at finite distance from the center of a strain bubble, it may be termed a finite strain bubble. On the other hand if Σ extends to infinity, then the strain bubble may be termed as an infinite strain bubble.
- ***Type or Mode of Strain Wave Oscillations.*** The strain wave oscillations sustained within a boundary surface Σ, may be either a standing wave type or a propagating phase wave type. The standing wave oscillations can only occur along one or two coordinate directions, within a finite 'core' region of any strain bubble. The propagating phase wave type oscillations along one of the coordinate directions, may be sustained within an infinite field of any strain bubble, with a decaying amplitude. However, the total strain energy content stored even in an infinite field must remain finite and constant. In some situations, propagating phase wave oscillations may be set up within a cylindrical ring type boundary surface, along ϕ coordinate direction, giving rise to 'spinning wave strain bubble'.

Strain Bubble Formation. If a certain finite amount of energy is somehow transferred to a particular region of the Elastic Space Continuum such that it attains a stable configuration in the deformed region, then a sustained strain field will develop in that region. The strain field within this particular region, called the 'strain bubble', will be completely defined by the displacement vector components u^i obtained from the solution of equilibrium equations (17.1) subject to the boundary conditions characterizing the physical situation. One of the crucial conditions for the formation and stability of such strain bubbles is the time invariance or conservation of the total strain energy contained in the strain field. Although the strain components will always be functions of space & time coordinates, yet the strain energy density may or may not be time invariant. A further condition for the stability of strain bubbles is the time invariance of its strain energy density. Even with such constraints, a large number of different varieties of strain bubbles can exist or coexist within the Elastic Space Continuum. Further, all strain bubbles experience characteristic interactions among themselves.

Strain Bubble Interactions & Potential Energy. If the strain fields of two strain bubbles overlap in a certain region of the Space Continuum, the total strain components will be obtained by superposing the corresponding components of both the strain bubbles referred to a common coordinate system. Strain components can be transformed from one coordinate system to another as per the rules for transformation of mixed tensor components. For example, if we have to transform strain tensor components $\varepsilon^i_j(x)$ defined in coordinate system (x^i) to strain tensor components $S^i_j(y)$ in coordinate system (y^i), we first need the coordinate transformation relations of the type,

$$y^i = f^i(x^1,x^2,x^3) \quad \& \quad x^i = F^i(y^1,y^2,y^3) \tag{17.2}$$

From these transformation relations we can obtain the Jacobean matrices of their partial derivatives $[\partial y^i/\partial x^j]$ and $[\partial x^i/\partial y^j]$. The required strain tensor components can now be obtained by using the relation,

$$S^i_j(y) = \frac{\partial y^i}{\partial x^\alpha} \cdot \varepsilon^\alpha_\beta \cdot \frac{\partial x^\beta}{\partial y^j} \tag{17.3}$$

Strain energy density and hence the total energy of the common field will be governed by the sum of squares of the resultant strain components. Interaction energy (E_{int}) of two such interacting strain bubbles may be defined as the difference between the total strain energy of the two strain bubbles with superposed strain fields (E_{sup}) and the sum of separate strain field energies of two bubbles (E_1 and E_2).

$$E_{int} = E_{sup} - (E_1 + E_2) \tag{17.4}$$

If $S^i_j(1)$ represents the strain components of strain bubble 1, and $S^i_j(2)$ represents the corresponding strain components of bubble 2, referred to the same coordinate system, then it can be easily seen from equation

(17.4) that the interaction energy density W_{int} will be given by the sum of products of the corresponding strain components as,

$$W_{int}(1,2) = (1/2\varepsilon_0).\Sigma[\{ S^i_j(1) + S^i_j(2) \}^2 - \{ S^i_j(1) \}^2 - \{ S^i_j(2) \}^2]$$

$$= (1/\varepsilon_0)\, \Sigma[S^i_j(1).\, S^i_j(2)] \quad (i \rightarrow 1 \text{ to } 3 \; \& \; j \rightarrow 1 \text{ to } 4) \qquad (17.5)$$

Similarly, the interaction energy density in the common overlapped region of more than two strain bubbles can be easily shown to be the sum of interaction energies of each pair of interacting strain bubbles as,

$$W_{int}(1,2,3) = (1/\varepsilon_0)\, \Sigma[\{S^i_j(1).\, S^i_j(2)\} + \{ S^i_j(1).\, S^i_j(3)\} + \{ S^i_j(2).\, S^i_j(3)\}] \,. \quad ... \qquad (17.6)$$

A negative interaction energy will imply the release of a portion of the total strain energy of the two interacting bubbles. The released energy will either transform into another strain bubble or wave packet, or transform into kinetic energy of motion of the interacting strain bubbles. In the extreme case of complete interaction between two strain bubbles with identical strain wave oscillations in opposite phase, the E_{sup} will reduce to zero and both strain bubbles may get annihilated with the released interaction energy transforming into one or more new strain bubbles or strain wave packets.

When the cores of two or more interacting strain bubbles get partly overlapped, the resulting interaction may be called 'core interaction' which is identical to the conventional strong interaction encountered among nucleons and other elementary particles. However, when the centers of interacting strain bubbles are so far apart as to preclude the core interactions, their propagating phase wave fields, if any, may still get superposed resulting in a wave field interaction.

The interaction energy of a pair of mutually interacting strain bubbles may be identified with the conventional potential energy. Thus, in the case of a +ve potential energy, external work has to be done or energy has to be supplied to the system from outside to account for the increase in the combined or superposed strain field energy (E_{sup}). On the other hand, in the case of -ve potential energy, a portion (E_{int}) of the total strain energy of the two bubbles is released from the overlapped/common strain field, which is either transformed into the kinetic energy of the interacting strain bubbles or emitted out of the system as a new strain bubble or strain wave packet.

Mutual attraction of two interacting strain bubbles can be easily attributed to their -ve interaction energy, the magnitude of which increases with reduction in their separation distance. Similarly, mutual repulsion of two interacting strain bubbles can be attributed to their +ve interaction energy. The wave field interactions, with negative interaction energy, between different pure or composite strain bubbles located quite far apart, will result in mutually 'bound' 'clusters' of strain bubbles. Formation of composite strain bubbles through core interactions with negative interaction energy and development of mutually bound clusters of various strain bubbles, is a significant phenomenon in the evolution of matter within the Elastic Space Continuum. The conventional material particles may be viewed at ultra-microscopic scale as clusters of various composite and pure strain bubbles bound together through mutual interactions.

Strain Bubbles & Elementary Particles. At a subatomic scale the primary constituents of matter, namely the electrons and nuclear particles, are known to occupy an extremely small volume fraction, of the order of 10^{-12} percent of the physical volume of any material body. The remaining bulk of intervening space is supposed to be empty or so called 'vacuum' with some electromagnetic fields 'existing' in this empty space. These 'material particles' concentrated in such a small volume fraction of entire space are essentially characterized by their 'mass', 'charge' and interaction properties. In the parlance of strain bubbles existing in the Elastic Space Continuum, the clusters of pure and composite strain bubbles depicting 'elementary particles' are essentially characterized by their 'strain energy content', 'phase wave or strain wave fields', if any, and their interaction properties.

In principle, there could be an infinitely large number of different types of strain bubbles occurring in the Elastic Space Continuum, that may be correlated with equally large number of stable and unstable elementary particles. Therefore, we must undertake detailed studies of various 'pure' and 'composite' strain bubbles occurring, forming or transforming, interacting and decaying in the Elastic Space Continuum.

17.3 Typical Solutions Representing Strain Bubbles

Cylindrically Symmetric Strain Bubbles. In cylindrical coordinates with physical components u^ρ, u^ϕ, u^z of displacement vector **U**, the equilibrium equations (17.1) take on the form of three partial differential equations as follows,

$$\frac{\partial^2 u^\rho}{\partial \rho^2}+\frac{1}{\rho}.\frac{\partial u^\rho}{\partial \rho}-\frac{u^\rho}{\rho^2}+\frac{1}{\rho^2}.\frac{\partial^2 u^\rho}{\partial \phi^2}+\frac{\partial^2 u^\rho}{\partial z^2}-\frac{2}{\rho^2}.\frac{\partial u^\phi}{\partial \phi}=\frac{1}{c^2}.\frac{\partial^2 u^\rho}{\partial t^2} \qquad (17.7A)$$

$$\frac{\partial^2 u^\phi}{\partial \rho^2}+\frac{1}{\rho}.\frac{\partial u^\phi}{\partial \rho}-\frac{u^\phi}{\rho^2}+\frac{1}{\rho^2}.\frac{\partial^2 u^\phi}{\partial \phi^2}+\frac{\partial^2 u^\phi}{\partial z^2}+\frac{2}{\rho^2}.\frac{\partial u^\rho}{\partial \phi}=\frac{1}{c^2}.\frac{\partial^2 u^\phi}{\partial t^2} \qquad (17.7B)$$

$$\frac{\partial^2 u^z}{\partial \rho^2}+\frac{1}{\rho}.\frac{\partial u^z}{\partial \rho}+\frac{1}{\rho^2}.\frac{\partial^2 u^z}{\partial \phi^2}+\frac{\partial^2 u^z}{\partial z^2} = \frac{1}{c^2}.\frac{\partial^2 u^z}{\partial t^2} \tag{17.7C}$$

Here, equations (17.7A) and (17.7B) are mutually coupled in ϕ coordinate. It is extremely difficult to obtain a general solution of these equations. However, it is important to note that any general solution of the above equations is not expected to satisfy all boundary conditions, especially the ones requiring the strain energy density and the total energy content to remain constant or time invariant. Hence, we look for some special solutions involving sinusoidal terms in time coordinate, that may yield time invariant energy density when the squares of strain components are summed up. A few examples of typical solutions of equilibrium equations (17.7) that satisfy the required boundary conditions and represent some of the 'pure' strain bubbles with cylindrical symmetry, are given below:

A. Stable Oscillating Core Strain Bubble. In accordance with earlier discussions of boundary conditions, one most important solution of equilibrium equations (17.7), that is independent of ϕ coordinate, is

$$u^\rho = \pm A_1.e\kappa.\ J_1(x).\ Cos(qz).\ Cos(\kappa ct) \tag{17.8A}$$

$$u^\phi = \pm A_1.e\kappa.\ J_1(x).\ Cos(qz).\ Sin(\kappa ct) \tag{17.8B}$$

$$u^z = 0 \tag{17.8C}$$

where A_1 is a dimensionless number, κ and q are dimensional constants with $x=(\kappa^2-q^2)^{1/2}\rho$, and $J_1(x)$ is the Bessel function of first order. The boundary surface Σ is given by, $-\pi/2 \le qz \le \pi/2$ and $0 \le x \le \alpha_1$ with $J_1(\alpha_1) = 0$ or $\alpha_1=3.832$. Here κ is the wave number of strain wave oscillations. The strain energy density W_1 for this strain bubble works out to be,

$$W_1 = \frac{A_1^2 e^2 \kappa^2}{2\varepsilon_0}\left[\left(\kappa^2-q^2\right)\left\{\left(J_1'(x)\right)^2+\frac{J_1^2(x)}{x^2}\right\}Cos^2(qz)\right] + \frac{A_1^2 e^2 \kappa^2}{2\varepsilon_0}\left[J_1^2(x)\left(\kappa^2 Cos^2(qz)+q^2 Sin^2(qz)\right)\right]$$

Since this energy density is completely independent of time, the strain bubble represented by equations (17.8) is expected to be most stable and will be identified later with the nucleon core. After integrating

W_l over the whole volume, the total strain energy E_1 of this strain bubble works out to be,

$$E_1 = \frac{\pi^2 A_1^2 e^2 \kappa \alpha_1^2 J_0^2(\alpha_1)}{2\varepsilon_0 \left(\frac{q}{k}\right)\left(1 - \frac{q^2}{k^2}\right)} \qquad (17.9)$$

The above expression for E_1 is minimized for $q = \kappa/\sqrt{3}$. This strain bubble displays very strong radial as well axial interactions. At any point $P(\rho,\phi,z)$ within the strain field of this bubble, the displacement vector **U** can be 'seen' to be rotating at constant angular velocity κc and with constant magnitude. This rotational motion of displacement vector **U** may be visualized as an intrinsic 'spin' of the strain field. The strong interactions of this strain bubble will be sensitive to the direction of this intrinsic spin vector relative to the 'spin' direction of the other interacting bubble.

B. Unstable Oscillating Core Strain Bubbles. Three important solutions in this category are:

$u^\rho = \pm A_2.e\kappa.\ J_1(x).\ \mathrm{Cos}(qz).\ \mathrm{Cos}(\kappa ct)$ with $u^\phi = 0$ & $u^z = 0$ (17.10)

$u^\phi = \pm A_3.e\kappa.\ J_1(x).\ \mathrm{Cos}(qz).\ \mathrm{Sin}(\kappa ct)$ with $u^\rho = 0$ & $u^z = 0$ (17.11)

and $u^z = \pm A_4.e\kappa.\ J_0(x).\ \mathrm{Cos}(qz).\ \mathrm{Sin}(\kappa ct)$ with $u^\rho = 0$ & $u^z = 0$ (17.12)

where, A_2, A_3, A_4 are dimensionless numbers, κ and q are dimensional constants with $x = (\kappa^2 - q^2)^{1/2}\rho$, and $J_0(x)$, $J_1(x)$ are the Bessel functions. The boundary surface Σ is given by $-\pi/2 \leq qz \leq \pi/2$ & $0 \leq x \leq \alpha_1$ with $J_1(\alpha_1) = 0$ for equations (17.10) and (17.11) and $J_0(\alpha_1) = 0$ for equation (17.12).

The strain energy density in these strain bubbles oscillates with time, thus rendering them unstable, even though the total strain energy remains time invariant. These strain bubbles are capable of strong interactions with other strain bubbles containing similar displacement vector components u^i. From the detailed study of their interactions, these strain bubbles are likely to be identified with the 'cores' of different mesons.

C. Spinning Wave Strain Bubbles. Another important solution in cylindrical coordinates represents a strain wave 'spinning' or going round and round in a cylindrical ring shaped region Σ. That is:

$u^\rho = \pm A_m.e\kappa.\ J_m(x).\ \mathrm{Sin}\{(m+1)\phi \pm \kappa ct\}.\ \mathrm{Cos}(qz)$ (17.13A)

$u^\phi = \pm A_m.e\kappa.\ J_m(x).\ \mathrm{Cos}\{(m+1)\phi \pm \kappa ct\}.\ \mathrm{Cos}(qz)$ (17.13B)

and $u^z = 0$ for $m \geq 1$; $x = (\kappa^2 - q^2)^{1/2}\rho$;
with $-\pi/2 \leq qz \leq \pi/2$ and $\alpha_n \leq x \leq \alpha_{n+1}$ for $J_m(\alpha_n) = 0$

In view of previous observations regarding phase wave fields, the strain energy density in this bubble is expected to be time invariant, thus rendering it a stable configuration. After detailed study of their interaction characteristics, this type of strain bubbles are likely to be used in major futuristic applications.

D. Spiral Wave Strain Bubbles. Another almost similar solution for u^z (with $u^\rho = u^\phi = 0$) consists of a strain wave spiraling along the Z-axis. This type of strain bubble is likely to have negligible interaction with other strain bubbles and may represent certain neutrino type particles.

$$u^z = \pm A_m.e\kappa.\ J_m(x).\ \text{Cos}(m\phi + qz \pm \kappa ct) \qquad (17.14)$$
$$\text{for } m \geq 1 \text{ and } x = (\kappa^2 - q^2)^{1/2}\rho\ ;$$

The boundary surface Σ for this strain bubble is given by,
$-\pi/2 \leq (m\phi + qz \pm \kappa ct) \leq \pi/2$ and $0 \leq x \leq \alpha_1$ with $J_m(\alpha_1) = 0$.

Spherically Symmetric Strain Bubbles. In spherical polar coordinates with physical components u^r, u^θ, u^ϕ of displacement vector **U**, the equilibrium equations (17.1) take on the form of three mutually coupled partial differential equations as follows,

$$\frac{\partial^2 u^r}{\partial r^2} + \frac{2}{r}.\frac{\partial u^r}{\partial r} - \frac{2}{r^2}.u^r + \frac{1}{r^2}\left(\frac{\partial^2 u^r}{\partial\theta^2} + \cot\theta.\frac{\partial u^r}{\partial\theta} + \frac{1}{\sin^2\theta}.\frac{\partial^2 u^r}{\partial\phi^2}\right)$$
$$-\frac{2}{r^2}\left(\frac{\partial u^\theta}{\partial\theta} + \cot\theta.u^\theta + \frac{1}{\sin\theta}.\frac{\partial u^\phi}{\partial\phi}\right) = \frac{1}{c^2}.\frac{\partial^2 u^r}{\partial t^2}$$

$$\frac{\partial^2 u^\theta}{\partial r^2} + \frac{2}{r}.\frac{\partial u^\theta}{\partial r} - \frac{u^\theta}{r^2.\sin^2\theta} + \frac{1}{r^2}\left(\frac{\partial^2 u^\theta}{\partial\theta^2} + \cot\theta.\frac{\partial u^\theta}{\partial\theta} + \frac{1}{\sin^2\theta}.\frac{\partial^2 u^\theta}{\partial\phi^2}\right)$$
$$+\frac{2}{r^2}\left(\frac{\partial u^r}{\partial\theta} - \frac{\cot\theta}{\sin\theta}.\frac{\partial u^\phi}{\partial\phi}\right) = \frac{1}{c^2}.\frac{\partial^2 u^\theta}{\partial t^2}$$

$$\frac{\partial^2 u^\phi}{\partial r^2} + \frac{2}{r}.\frac{\partial u^\phi}{\partial r} - \frac{u^\phi}{r^2.\sin^2\theta} + \frac{1}{r^2}\left(\frac{\partial^2 u^\phi}{\partial\theta^2} + \cot\theta.\frac{\partial u^\phi}{\partial\theta} + \frac{1}{\sin^2\theta}.\frac{\partial^2 u^\phi}{\partial\phi^2}\right)$$
$$+\frac{2}{r^2.\sin\theta}\left(\frac{\partial u^r}{\partial\phi} + \cot\theta.\frac{\partial u^\theta}{\partial\phi}\right) = \frac{1}{c^2}.\frac{\partial^2 u^\phi}{\partial t^2}$$

$$\ldots \qquad (17.15)$$

Here again, these equations are mutually coupled and hence it is extremely difficult to obtain any general solution of these equations. Therefore, we look for some special solutions involving sinusoidal terms in time coordinate, that may yield time invariant energy density. For a spherically symmetric boundary surface Σ, a few important solutions of equilibrium equations (17.15) for displacement components u^r, u^θ, u^ϕ, obtained by the method of separation of variables are given below. [10]

(a) Oscillating Core Strain Bubble. One lowest order solution of the equilibrium equations (17.15), under special constraint of null u^θ, with u^ϕ being independent of ϕ coordinate and u^r being independent of both θ and ϕ coordinates,is given by,

$$u^r = -A_e.e\kappa.(\pi/2x)^{1/2}.\ J_{1+1/2}(x).Cos(\kappa ct)$$
$$=A_e.e\kappa.G_1(x).\ Cos(\kappa ct) \qquad (17.16A)$$

$$u^\phi = A_e.e\kappa.(\pi/2x)^{1/2}.J_{1+1/2}(x).Sin(\theta).Sin(\kappa ct)$$
$$= -\ A_e.e\kappa.G_1(x).Sin(\theta).Sin(\kappa ct) \qquad (17.16B)$$

and $u^\theta = 0;$ (17.16C)

where, $G_1(x) = -(\pi/2x)^{1/2}.\ J_{1+1/2}(x) = [Cos(x)-Sin(x)/x]/x$, with $x = \kappa\, r$.

The radial boundary surface Σ for this strain bubble is given by,

$$0 \le x \le a_1 \qquad \text{with} \quad J_{1+1/2}(a_1) = 0 \quad \text{or} \quad a_1 = 4.4934$$

The strain energy density W_e for this strain bubble, computed from the strain components works out to be,

$$W_e = \frac{A_e^2 e^2 \kappa^4}{2\varepsilon_0}.\left\{\left(G_1'(x)\right)^2 + \frac{2G_1^2(x)}{x^2} + G_1^2(x).Sin^2(\theta)\right\}.Cos^2(\kappa ct)$$
$$+\frac{A_e^2 e^2 \kappa^4}{2\varepsilon_0}.\left\{G_1^2(x) + \frac{G_1^2(x)}{x^2}.\left(Cos^2(\theta)+1\right) + \left(G_1'(x)\right)^2.Sin^2(\theta)\right\}.Sin^2(\kappa ct)$$

which is not invariant with time, thus indicating long term instability of this strain bubble. Further, the total strain energy of this bubble, computed by integrating W_e over the whole volume, works out to be $E_e=7.1356\ \pi A_e^2 e^2\kappa/\varepsilon_0$. However, this oscillating core can degenerate into a lower energy state consisting of a part of this oscillating core surrounded by a radial phase wave or a strain wave field.

(b) Strain Bubble with Radial Wave Field. For this strain bubble, the displacement vector components for the core region are, from equation (17.16) ,

$u^r = A_e.e\kappa.G_l(x).\ Cos(\kappa ct);$

$u^\phi = -\ A_e.e\kappa.G_l(x).\ Sin(\theta).\ Sin(\kappa ct);$

and $u^\theta = 0$ with $0 \leq x \leq b_l$ where $x = \kappa\, r$ and the magnitude of parameter b_l is determined from the zero of Bessel function $J_{-l-½}(b_l) = 0$. Further, let $H_l(x) = -(\pi/2x)^{½}.\ J_{-l-½}(x) = [Sin(x)+Cos(x)/x]/x$.

For the wave field region $x \geq b_l$ let us consider another solution of equilibrium equations (17.15) consisting of a combination of $G_l(x)$ and $H_l(x)$ functions as follows:

$$u^r = A_e.e\kappa.\{G_l(x).\ Cos(\kappa ct) - H_l(x).\ Sin(\kappa ct)\}$$
$$= A_e.e\kappa.G_l(x,\psi_-)$$
$$\approx (A_e.e\kappa/x)\ .Cos(\psi_-) \qquad (17.17A)$$

$$u^\phi = -\ A_e.e\kappa.\{G_l(x).Sin(\kappa ct) + H_l(x).\ Cos(\kappa ct)\}.\ Sin(\theta)$$
$$= -A_e.e\kappa.H_l(x,\psi_-).\ Sin(\theta)$$
$$\approx -(A_e.e\kappa/x).\ Sin(\theta).Sin(\psi_-) \qquad (17.17B)$$

$$u^\theta = 0\ ; \qquad (17.17C)$$

here, $\psi_- = x+\kappa ct$; $G_l(x,\psi_-) = [Cos(\psi_-) - Sin(\psi_-)/x]/x$;

and $H_l(x,\psi_-) = [Sin(\psi_-) + Cos(\psi_-)/x]/x$.

The strain energy density W_e for the core region is still the same as given above, but the total strain energy for the whole bubble has now decreased to $E_e = 5.04\pi A_e^2 e^2\kappa/\varepsilon_0$. The reduction in this total energy, and spread of a part of its strain energy into the radial phase wave or strain wave field renders the bubble an inherent stability even though the strain energy density still oscillates slightly. This strain bubble can be identified with the elementary particle electron and the radial strain wave field is expected to represent the electrostatic field of charge particles. The radial direction of propagation of phase waves in this solution distinguishes between the fields of electron and positron.

Due to the phase wave considerations discussed above, the radial strain wave field of this bubble behaves like an A.C. voltage and the effective strain components in this field are given by the rms values of their peak magnitudes. At large distances, the $1/x^2$ terms may be neglected in comparison with $1/x$. The interaction energy of two overlapping 'strain wave' fields can then be computed easily to verify the Coulomb interaction law. Since the u^r and u^ϕ components here are in quadrature to each other, the intrinsic spin occurs in this strain wave field also. The u^r and u^ϕ components in the core will be in phase opposition for the electron and positron.

(c) Spinning Wave Core Strain Bubbles. Another important class of solutions of equilibrium equations (17.15) is obtained with special constraint of null u^ϕ and both u^r as well as u^θ being sinusoidal functions of ϕ and t coordinates. One such specially constrained solution consists of spinning wave core type strain bubbles represented by u^r and u^θ as given below,

$$u^r = \pm A_{e1}.e\kappa.G_1(x).\ \mathrm{Sin}(\theta)\mathrm{Cos}(\theta).\mathrm{Cos}(\phi \pm \kappa ct). \qquad (17.18A)$$

$$u^\theta = \pm A_{e1}.e\kappa.G_1(x).\ \mathrm{Sin}^2(\theta).\mathrm{Cos}(\phi \pm \kappa ct). \qquad (17.18B)$$

$$u^\phi = 0\ ; \qquad (17.18C)$$

where $x = \kappa$ r and $0 \le x \le a_1$ with $J_{1+½}(a_1) = 0$ or $a_1 = 4.4934$

This strain bubble too is expected to be inherently stable and after studying its interaction characteristics, may be identified with a neutrino type particle.

17.4 Kinetic Energy of Strain Bubbles and QM

The total strain energy stored in any strain bubble at 'rest' in the Elastic Space Continuum, may be treated as its rest mass energy or 'bound energy'. Apart from the change in their total 'bound energy' during interaction of two strain bubbles, the magnitude of dynamic stresses in their common region may either increase (positive interaction energy) or decrease (negative interaction energy), thereby disturbing the symmetric distribution of dynamic stresses in both strain bubbles. As a result of this asymmetry induced in dynamic stress field during interaction, equal and opposite resultant forces F_{int} will start acting on both strain bubbles, tending to move them in such a way as to reduce their total bound energy.

The motion of interacting strain bubbles may be visualized as the motion of their respective 'center of mass' points. With the motion of each strain bubble possessing non-zero rest mass, we associate the terms kinetic energy and momentum as per their conventional definitions. As mentioned earlier, the negative interaction energy of interacting strain bubbles is the amount of energy released from their 'bound' or 'mass' energies during interaction and gets transferred to the kinetic energies of their motion in accordance with the laws of conservation of energy & momentum. The exact mechanism of this transfer of interaction energy to the kinetic energy is expected to be quite a complex phenomenon and needs to be investigated separately.

The most pertinent point here is that just as all other forms of energy exist in the Elastic Space Continuum as strain energy of various strain bubbles, the kinetic energy associated with the motion of any strain bubble also must exist in some sort of 'strain wave field' associated with the motion of that strain bubble. But we know from Quantum Mechanics that the only waves of non-electromagnetic origin, associated with the motion of microscopic particles, are the de Broglie waves represented by 'ψ' wave function. Hence, logically, the strain wave field associated with the motion of a strain bubble, must be identified with the 'ψ wave field' associated with the motion of that strain bubble. However, the only waves of non-electromagnetic origin that could be induced in the Elastic Space Continuum, are the longitudinal strain waves that must therefore be identified with the 'ψ wave field'. Now, we may visualize the uniform motion of a strain bubble as a state in which a moving 'ψ wave field', carrying a definite amount of total strain energy (i.e. kinetic energy), is induced or associated with the strain bubble in motion.

Therefore, the change in motion of a strain bubble may be visualized as a process or phenomenon during which the interaction energy gets transferred to the kinetic energy or 'total strain energy of the associated ψ wave field' and vice versa. Since the bubble interactions and such energy transfer processes are limited by finite velocity of light 'c' due to their inherent 'spatial spread', classical mechanics may be considered adequate for describing the motion of strain bubbles at low velocities. However, at higher velocities and corresponding high energy interactions, adequate study and analysis of the associated phenomenon can only be made by using the techniques of Wave Mechanics. But the fundamental concepts of Wave Mechanics may have to be thoroughly revised and refined in the light of present Space Dynamics studies. Most importantly, the interpretation of ψ wave field needs to be changed from the current 'probability wave' to the pilot wave representing the kinetic energy and momentum of the strain bubble in motion, some what in line with the Bohmian Quantum Mechanics.

Summary and Conclusion. We started with a pragmatic observation that our familiar vacuum or free space continuum with the characteristic property of permittivity ε_0 and permeability μ_0 behaves as a perfect isotropic elastic continuum with elastic constant $1/\varepsilon_0$ and inertial constant μ_0. We have then given a detailed description of displacement vector **U**, strain tensor **S** and stress tensor **T** in this space continuum. Precluding 'atomicity' and rigid body motions in the 'Elastic Space Continuum', we have used a simple modified form of Hooke's law and derived ab-initio the dynamic equilibrium equations of elasticity. These equilibrium equations are found to be identical with the vector wave equation of Maxwell's electromagnetic theory. Particular solutions of these equilibrium equations, as functions of space and time coordinates and satisfying appropriate boundary and stability conditions, are shown to represent various 'strain bubbles' and 'strain wave fields'.

The electromagnetic field as well as all other forms of particles, are shown to exist in the Elastic Space Continuum as strain wave fields or strain bubbles with definite amount of strain energy associated with them. Mutual interactions among various strain bubbles and fields are shown to be governed by the increase or decrease in strain intensity in their common superposed strain field. The clusters of pure and composite strain bubbles depicting 'material particles' are essentially characterized by their strain energy content, phase wave or strain wave fields and their interaction properties. Therefore, it is imperative that for deeper insight and a more fundamental understanding of elementary and composite material particles and the associated phenomenon at an ultra microscopic level, we must undertake detailed studies of corresponding strain bubbles occurring, transforming, moving, interacting and decaying in the Elastic Space Continuum.

18

The Electron Structure and Coulomb Interaction

18.1 Spherically Symmetric Strain Bubbles

Introduction. Of the thousands of elementary particles known so far, the electron was the first one to be detected, most actively researched and studied. It has been used in many applications and is the most well known of all elementary particles. As a part of our understanding about the electron, we have accurately determined its charge, mass, spin, angular momentum, magnetic moment and interaction characteristics. Apparently, everything that is worth knowing about the electron is already known. Yet, most scientists still regard the electron as a point charge, a point mass and a structure-less elementary particle. We are still not in a position to know or visualize the shape, size and inner structural details of the electron.

Ideally speaking, the mental visualization of a physical situation must precede the use of mathematical techniques for its logical analysis. However, we can not develop a very clear and vivid mental picture of a physical situation as long as the fundamental concepts associated with that situation are themselves vague and hazy. For instance, in the present case we can't develop a very clear mental picture of Coulomb interaction between two electrons as long as the fundamental concepts of electron structure, its charge property and its electrostatic field are themselves not clear. Hence, based on the elastic properties of space continuum, we will first examine some of the spherically symmetric solutions of equilibrium equations of elasticity in the Continuum to derive the structure of electron and its electrostatic wave field. We shall then develop a model to compute the interaction energy from the superposition of effective strain wave fields of two electrons separated by a finite distance to verify Coulomb's interaction law and deduce the concept of charge property.

Electrostatic Wave Field. It is known that an electrostatic field influence propagates or spreads out from the source particle at the velocity of light. It is also believed that electromagnetic interaction is mediated through continuous exchange of 'virtual photons' between the charge particles. Even a 'virtual photon' field is often assumed to be surrounding all charge particles to account for the simultaneous Coulomb interaction among infinitely many such particles. Hence, the electrostatic field seems to be inherently dynamic instead of being static in character. That is, instead of being a function f(R) of relative position vector **R** alone, it may have to depend on time as well, so that the field influence appears to be spreading at velocity of light. Therefore, we may characterize the electrostatic wave field by a function of the type $f(R).e^{i\kappa(R-ct)}$, where $i = (-1)^{1/2}$, representing a sort of spherical phase wave of amplitude f(R) propagating radially at velocity of light c. Here, κ represents the wave number. This hazy picture of the electrostatic field is further developed through solutions of equilibrium equations in the elastic space continuum.

The equilibrium equations of elasticity written in terms of displacement vector **U** in the elastic space continuum turn out to be identical to the vector wave equation. These equations in vector and tensor form are given below,

$$\partial^2\mathbf{U}/\partial x^2 + \partial^2\mathbf{U}/\partial y^2 + \partial^2\mathbf{U}/\partial z^2 = \nabla^2\mathbf{U} = (1/c^2)\,\partial^2\mathbf{U}/\partial t^2 \qquad (18.1A)$$

$$g^{11}u^i_{,11} + g^{22}u^i_{,22} + g^{33}u^i_{,33} = g^{jj}u^i_{,jj} = (1/c^2)\,\partial^2 u^i/\partial t^2 \qquad (18.1B)$$

where, the displacement vector components u^i are continuous functions of space and time coordinates (y^1, y^2, y^3, t). As shown in chapter 16, the following correlation exists between displacement vector field **U** and the electromagnetic field vectors **E** and **B** when $\nabla.\mathbf{U} = 0$.

$$\mathbf{E} = -\,(1/\varepsilon_0).(1/c).\partial\mathbf{U}/\partial t \qquad (18.2)$$

$$\mathbf{B} = (1/c).(1/\varepsilon_0).\,(\nabla\times\mathbf{U}) \qquad (18.3)$$

An electromagnetic field in free space is found to be a dynamic stress-strain field in the Elastic Space Continuum. To study the strain field structure of the electron we need to study the spherically symmetric solutions of equilibrium equations (18.1).

Spherically Symmetric Solution. The complete electron structure will now be shown as a special solution of equilibrium equations (18.1) in spherical polar coordinate system (y^1=R, y^2=θ, y^3=ϕ). If we write equation (18.1) in terms of physical components u^R, u^θ and u^ϕ of displacement vector **U** in spherical coordinate system, we get a set of three simultaneous partial differential equations (17.15), the general solution of which is most intricate due to the mutual coupling of these displacement components. One of the lowest order solutions of these equations is obtained when we restrict u^θ=0, u^ϕ to be independent of ϕ coordinate and u^R to be independent of both θ and ϕ coordinates. The resulting equations reduce to,

$$\frac{\partial^2 u^R}{\partial R^2} + \frac{2}{R}.\frac{\partial u^R}{\partial R} - \frac{2}{R^2}.u^R = \frac{1}{c^2}.\frac{\partial^2 u^R}{\partial t^2} \qquad (18.4)$$

$$\frac{\partial^2 u^\phi}{\partial R^2} + \frac{2}{R}.\frac{\partial u^\phi}{\partial R} - \frac{1}{R^2\sin^2\theta}.u^\phi + \frac{1}{R^2}.\left(\frac{\partial^2 u^\phi}{\partial\theta^2} + \cot\theta.\frac{\partial u^\phi}{\partial\theta}\right) = \frac{1}{c^2}.\frac{\partial^2 u^\phi}{\partial t^2}$$

... (18.5)

18.2 The Electron Structure

The Standing Wave Core. The oscillating wave type solution of the above equations (4) and (5) with a spherically symmetric boundary surface, is given for positron (+ve) and electron (-ve) cores by:

$$u^R = \pm A_e.e\kappa.G_1(X). \mathrm{Cos}(\kappa ct); \qquad (18.6A)$$

$$u^\phi = \pm A_e.e\kappa.G_1(X). \mathrm{Sin}(\theta). \mathrm{Sin}(\kappa ct); \qquad (18.6B)$$

where, $G_1(X) = (\cos X - \sin X / X)/X = -(\pi/2x)^{1/2}. J_{3/2}(X)$ and $X = \kappa R$.

Equations (18.6) represent the standing strain wave type oscillations within the core region. Another similar solution which has a singularity at the origin and hence not admissible for the electron/positron core is given by:

$$u^R = \pm A_e.e\kappa.H_1(X). \mathrm{Sin}(\kappa ct); \qquad (18.7A)$$

$$u^\phi = \pm A_e.e\kappa.H_1(X). \mathrm{Sin}\theta. \mathrm{Cos}(\kappa ct); \qquad (18.7B)$$

where, $H_1(X) = (\sin X + \cos X / X)/X = -(\pi/2x)^{1/2}. J_{-3/2}(X)$ and $X = \kappa R$,

and $b_1 \leq X \leq \infty$ with $J_{-1-1/2}(b_1) = 0$ & $b_1 = 2.7984$

The Radial Wave Field. However, if the above two solutions are combined together we get an oscillating core given by equations (18.6) for $0 \leq X \leq b_1$ and a propagating radial phase wave solution for the electron field for $X \geq b_1$ given by,

$$u^R = -A_e.e\kappa.\{G_1(X). \mathrm{Cos}(\kappa ct) - H_1(X). \mathrm{Sin}(\kappa ct)\}$$
$$= -A_e.e\kappa.G_1(X,\psi_-) \approx -(A_e.e\kappa/X).\mathrm{Cos}(\psi_-) \qquad (18.8A)$$

$$u^\phi = -A_e.e\kappa.\{G_1(X).\mathrm{Sin}(\kappa ct) + H_1(X). \mathrm{Cos}(\kappa ct)\}. \mathrm{Sin}(\theta)$$
$$= -A_e.e\kappa.H_1(X,\psi_-). \mathrm{Sin}(\theta) \approx -(A_e.e\kappa/X). \mathrm{Sin}(\theta).\mathrm{Sin}(\psi_-) \qquad (18.8B)$$

$u^\theta = 0$; where, $\psi_- = X + \kappa ct$

with, $G_1(X,\psi_-) = [\mathrm{Cos}(\psi_-) - \mathrm{Sin}(\psi_-)/X]/X$;

and $H_1(X,\psi_-) = [\mathrm{Sin}(\psi_-) + \mathrm{Cos}(\psi_-)/X]/X$.

This strain wave field consisting of phase waves propagating inwards from infinity to the core boundary, at the speed of light c, represents the electrostatic field of an electron type charge particle. Another similar solution consisting of phase waves propagating outwards from the core boundary to infinity, at the speed of light c, representing the electrostatic field of positron type charge particle is be given by,

$$u^R = A_e.e\kappa.\{G_1(X). \mathrm{Cos}(\kappa ct) + H_1(X). \mathrm{Sin}(\kappa ct)\}$$
$$= A_e.e\kappa.G_1(X,\psi_+)$$
$$\approx (A_e.e\kappa/X).\mathrm{Cos}(\psi_+) \text{ (for } X >> b_1) \qquad (18.9A)$$

$$u^{\phi} = A_e.e\kappa.\{G_1(X).Sin(\kappa ct)- H_1(X). Cos(\kappa ct)\}. Sin(\theta)$$
$$= -A_e.e\kappa.H_1(X,\psi_+). Sin(\theta)$$
$$\approx -(A_e.e\kappa/X). Sin(\theta).Sin(\psi_+) \quad (\text{for } X >> b_1) \tag{18.9B}$$

$u^{\theta} = 0$; where, $\psi_+ = X\text{-}\kappa ct$

with, $G_1(X,\psi_+) = [Cos(\psi_+)-Sin(\psi_+)/X]/X$;

and $H_1(X,\psi_+) =[Sin(\psi_+)+Cos(\psi_+)/X]/X$.

Physical Components of Strain. The strain energy of the 'core' and 'strain wave field' can now be computed from the corresponding strain components. Here, the wave number κ is of the order of 10^{15} m^{-1}, A_e is a dimensionless constant, and 'e' the magnitude of electron charge. The solutions u^R and u^{ϕ} are in phase quadrature in the core as well as in the field. The elements of strain tensor S^i_j can be computed by taking covariant space and time derivatives of displacement components u^i for spatial and temporal strain terms. The mixed tensor components S^i_j can be converted to the corresponding physical components by using the relation,

$$S^{y^i}_{y^j} = \sqrt{g_{ii}}\,.\,S^i_j\,.\,\sqrt{g^{jj}} \quad \text{(No summation over i or j)} \tag{18.10}$$

where g^{ij} are the usual metric tensor components for the reference coordinate system. The physical strain components of **S** and the corresponding strain energy density W can therefore be directly computed from the following relations:

$$S^R_R = \frac{\partial u^R}{\partial R} \;;\; S^R_\theta = \frac{1}{R}.\frac{\partial u^R}{\partial \theta} - \frac{u^\theta}{R} \;;\; S^R_\phi = \frac{1}{R.\sin\theta}.\frac{\partial u^R}{\partial \phi} - \frac{u^\phi}{R}$$

$$S^\theta_R = \frac{\partial u^\theta}{\partial R} \;;\; S^\theta_\theta = \frac{1}{R}.\frac{\partial u^\theta}{\partial \theta} + \frac{u^R}{R} \;;\; S^\theta_\phi = \frac{1}{R.\sin\theta}.\frac{\partial u^\theta}{\partial \phi} - \frac{\cot\theta}{R}.u^\phi$$

$$S^\phi_R = \frac{\partial u^\phi}{\partial R} \;;\; S^\phi_\theta = \frac{1}{R}.\frac{\partial u^\phi}{\partial \theta} \;;\; S^\phi_\phi = \frac{1}{R.\sin\theta}.\frac{\partial u^\phi}{\partial \phi} + \frac{\cot\theta}{R}.u^\theta + \frac{u^R}{R}$$

$$S^R_t = \frac{1}{c}.\frac{\partial u^R}{\partial t} \;;\; S^\theta_t = \frac{1}{c}.\frac{\partial u^\theta}{\partial t} \;;\; S^\phi_t = \frac{1}{c}.\frac{\partial u^\phi}{\partial t} \tag{18.11}$$

$$W = (1/2\varepsilon_0).\sum |S^{y^i}_{y^j}|^2 \;. \tag{18.12}$$

Core & Field Strain Components. Since, in the strain wave field, surfaces of constant phase propagate outwards or inwards at the speed of light 'c', without any associated transport of strain energy, we may also term this wave field as phase wave field. However, there is a special feature in this phase wave field. Whereas, in the standing or oscillating wave solutions the temporal and spatial strain components are in quadrature; in propagating phase wave field, they are in phase opposition for Cos($\psi_{\pm}$) and Sin($\psi_{\pm}$) terms and get canceled out. Hence, the energy density in the phase wave field will be governed only by the maximum amplitude of these phase waves or more precisely by their rms values. The physical strain components for the positron core, computed from relations (18.6) and (18.11) are listed below in terms of functions G_1 and H_1 defined above.

$$S_R^R=-A_e.e\kappa^2.G_1'(X).Cos(\kappa ct)\ ; \qquad S_\phi^R=A_e.e\kappa^2.\frac{G_1(X)}{X}.Sin(\theta).Sin(\kappa ct)$$

$$S_\phi^\phi=-A_e.e\kappa^2.\frac{G_1(X)}{X}.Cos(\kappa ct)\ ; \quad S_\theta^\phi=-A_e.e\kappa^2.\frac{G_1(X)}{X}.Cos(\theta).Sin(\kappa ct)$$

$$S_\phi^\theta=A_e.e\kappa^2.\frac{G_1(X)}{X}.Cos(\theta)Sin(\kappa ct)\ ; \qquad S_\theta^\theta=-A_e.e\kappa^2.\frac{G_1(X)}{X}.Cos(\kappa ct)$$

$$S_R^\phi=-A_e.e\kappa^2.G_1'(X).Sin(\theta).Sin(\kappa ct)\ ; \qquad S_t^R=A_e.e\kappa^2.G_1(X).Sin(\kappa ct)$$

$$S_t^\phi=-A_e.e\kappa^2.G_1(X).Sin(\theta).Cos(\kappa ct)\ ; \qquad (18.13)$$

where $G_1'(X) = -(Sin(X)+2.G_1(X))/X$

The strain components for the electron core will be just the same as given above but with opposite signs for all components. However, the strain components derived from u^ϕ component will experience sign change on reversal of the spin axis.

In the strain wave field of the electron represented by equations (18.8), the simplified strain components for large X, where $1/X^2$ terms could be neglected in comparison to 1/X terms, are given by,

$$S_\phi^R=\frac{A_e.e\kappa^2.Sin(\theta)}{X^2}.Sin(\psi_-)\ ; \quad S_R^\phi=\frac{A_e.e\kappa^2.Sin(\theta)}{X^2}.Sin(\psi_-)$$

$$S_\phi^\theta=\frac{A_e.e\kappa^2.Cos(\theta)}{X^2}.Sin(\psi_-)\ ; \quad S_\theta^\phi=-\frac{A_e.ek^2.Cos(\theta)}{X^2}.Sin(\psi_-)$$

$$S_R^R=\frac{A_e.e\kappa^2}{X^2}.Cos(\psi_-)\ ; \ S_\theta^\theta=S_\phi^\phi=-\frac{A_e.e\kappa^2}{X^2}.Cos(\psi_-) \qquad (18.14)$$

For a stationary particle the amplitude of these phase waves at any point will remain constant with time. Here too the strain components derived from u^ϕ component will experience sign change on reversal of the spin axis. This could imply that the Coulomb interaction of two charge particles may, to some extent, depend upon the orientation of their spin axes. However, we have not examined this aspect in any detail in this work.

Core & Field Strain Energy. The strain energy density W, calculated by using equations (18.12) to (18.14), is given below as W_c for the core and W_f for the field,

$$W_c = \frac{A_e^2 e^2 \kappa^4}{2\varepsilon_0} \cdot \left\{ \left(G_1'(x)\right)^2 + \frac{2G_1^2(x)}{x^2} + G_1^2(x).\,\mathrm{Sin}^2(\theta) \right\} .\mathrm{Cos}^2(\kappa\, ct)$$

$$+ \frac{A_e^2 e^2 \kappa^4}{2\varepsilon_0} \cdot \left\{ G_1^2(x) + \frac{G_1^2(x)}{x^2}.\left(\mathrm{Cos}^2(\theta)+1\right) + \left(G_1'(x)\right)^2.\,\mathrm{Sin}^2(\theta) \right\} .\,\mathrm{Sin}^2(\kappa\, ct)$$

... (18.15)

$$W_f = \frac{A_e^2 e^2 \kappa^4}{2\varepsilon_0} \left[\frac{3.\mathrm{Cos}^2(\psi\_)}{X^4} + \frac{2.\mathrm{Sin}^2(\psi\_)}{X^4} \right] \quad (18.16)$$

In W_c above, the coefficients of $\mathrm{Cos}^2(\kappa ct)$ and $\mathrm{Sin}^2(\kappa ct)$ terms are not exactly equal, which indicates energy density fluctuations within the core. Possibly, these minor fluctuations in the energy density as well as the total energy content in the core are accommodated through slight fluctuations in the core boundary during the period of each oscillation cycle. However, for the overall total energy computation, we may take the time averaged value of the energy density as,

$$W_c = \frac{A_e^2 e^2 k^4}{4\varepsilon_0} \cdot \left[\left(\left(G_1'(X)\right)^2 + G_1^2(X) \right).\left(1+\mathrm{Sin}^2(\theta)\right) + \frac{G_1^2(X)}{X^2}.\left(3+\mathrm{Cos}^2(\theta)\right) \right]$$

and $W_f = (1/4\varepsilon_0).\ A_e^2.e^2.\kappa^4[\ 5/X^4\ ]$ (18.17)

The total energy of the electron is given as a sum of the core and field energies obtained from the volume integral of W_c over the core region ($0 \le X \le b_1$) and W_f over the field region ($X \ge b_1$), as

$$E_{total} = E_c + E_f = (1/\varepsilon_0).\pi.\ A_e^2.e^2.\kappa.[\ 3.2533 + 1.7867\ ]$$
$$= (5.04/\varepsilon_0).\pi.\ A_e^2.e^2.\kappa. = \mathbf{m_e.c^2} \quad (18.18)$$

This shows that almost 65 percent of the total mass energy of the electron is contained in its core and remaining 35 percent in its field. In the above relation (18.18) where $\mathbf{m_e}$ is the known mass of the electron, there are two unknown parameters A_e and κ. For a unique determination of these parameters, we need one more relation which will be obtained from the Coulomb interaction model.

Intrinsic Spin Effect. As already indicated above, the only difference in the wave fields of positron (ψ_+) and electron (ψ_-) is in the opposite directions of propagation of their phase waves. In both cases, as seen from equations (18.6) to (18.9), the displacement components u^R and u^ϕ are in quadrature to each other. If we denote Z-axis as the axis of the electron or positron strain bubble, then all planes perpendicular to this axis may be referred as transverse planes. It can be easily seen from phase quadrature of the displacement vector components that the resultant displacement vector in any transverse plane keeps continuously rotating with constant angular velocity $\omega=\kappa c$, whereas its magnitude remains constant or time invariant at any space point. In the principal transverse plane ($\theta=\pi/2$), the magnitude of resultant displacement vector **U** in the wave field remains constant with $|\mathbf{U}| =\sqrt{2}.(A_e.e\ /R)$. Throughout this principal transverse plane, the constant magnitude vector **U** keeps rotating or 'spinning' with constant angular velocity $\omega=\kappa c$. The direction of this spin of the displacement vector is obviously along the axis of the strain bubble and remains constant with time.

This constant 'intrinsic spin' of the displacement vector found in the core as well as strain wave field of the electron/positron type strain bubbles may be identified with the conventional notion of 'spin' in these particles. The phenomenon of this intrinsic spin is a very unique feature in the ultra-microscopic realm of Nature. It can be seen from equations (18.6) to (18.9) that in normal orientation a positron will have a +ve direction of intrinsic spin, along +ve Z axis, and radially outward propagating phase wave field identified with +ve charge. If the axis of the positron is reversed, the direction of its spin will become -ve but the phase waves will still be propagating radially outwards. Similarly for an electron, in any orientation of the intrinsic spin, its phase waves will keep on propagating radially inwards.

18.3 Coulomb Interaction

Effective 'rms' Valued Strain Field Model. When the cores of two interacting strain bubbles get overlapped or superposed, their resulting strong interaction can be computed in a straight forward manner. Their actual strain components, referred to a common coordinate system, are directly superposed to compute the resulting interaction energy over the common overlapped region. However, computation of the interaction energy in the common overlapped region of strain wave fields of two charge particles, appears to be a complex problem due to the presence of phase waves. Even though these phase waves do not transport strain energy, yet they appear to display certain 'wave momentum' effects. In a way these strain waves could be compared with sinusoidal AC voltages. Because of the inherent phase opposition of strain components in the cores of positron and electron, leading to the opposite directions of propagation of their phase waves, the strain wave fields of the electron and positron will show inherent opposition such that when superposed they will tend to cancel out each other. The field energy density in each case is governed by the 'rms.' values of the amplitudes of their respective phase waves. We may therefore, adopt the rms value concept for the magnitude of the effective strain components in the field.

For the purpose of developing a simplified model of Coulomb interactions we consider only the rms valued amplitudes of the respective strain components. Accordingly, we assign $+(1/\sqrt{2})$ for the $\mathrm{Cos}(\psi_+)$ or $\mathrm{Sin}(\psi_+)$ terms and $-(1/\sqrt{2})$ for the $\mathrm{Cos}(\psi_-)$ or $\mathrm{Sin}(\psi_-)$ terms occurring in the strain components, corresponding to the phase waves propagating outwards from or inwards to the source particle. Corresponding to equations (18.14), we may specifically write out the effective strain components for electrostatic wave field of an electron as,

$$S_R^R = -\frac{1}{\sqrt{2}}\cdot\frac{A_e.e\kappa^2}{X^2}\ ; \qquad S_\theta^\theta = S_\phi^\phi = \frac{1}{\sqrt{2}}\cdot\frac{A_e.e\kappa^2}{X^2}$$

$$S_\phi^R = -\frac{1}{\sqrt{2}}\cdot\frac{A_e.e\kappa^2.\mathrm{Sin}(\theta)}{X^2}\ ;\ S_R^\phi = -\frac{1}{\sqrt{2}}\cdot\frac{A_e.e\kappa^2.\mathrm{Sin}(\theta)}{X^2}$$

$$S_\phi^\theta = -\frac{1}{\sqrt{2}}\cdot\frac{A_e.e\kappa^2.\mathrm{Cos}(\theta)}{X^2}\ ;\ S_\theta^\phi = \frac{1}{\sqrt{2}}\cdot\frac{A_e.e\kappa^2.\mathrm{Cos}(\theta)}{X^2} \qquad (18.19)$$

The corresponding effective strain components for a positron field will have opposite signs to the ones given above.

To compute the Coulomb interaction energy between two charge particles, [1,2], we have to superpose the effective strain tensor components S^i_j of their respective wave fields in a common coordinate system and then compute the total energy of their combined fields. This total field energy E_{sup} may be more or less than the sum of their isolated field energies. The difference $[E_{int} = E_{sup} - (E_1 + E_2)]$ is termed as interaction energy. For two similar charges the respective strain components add up and since the energy density is proportional to the sum of squares of strain components, their total field energy will be more than the sum of their separate field energies. Conventionally the interaction energy is termed +ve in this case. Similarly, for two dissimilar charges the combined field energy will reduce and the interaction energy will be termed -ve. The negative interaction energy implies that due to the superposition of fields, part of the initial total field energy of the system of interacting charges is released by the system and may get transformed to some other form. As a limiting case, when the separation between the two charge particles is reduced to zero, their interaction energy will not become infinity but will be limited to the sum of their initial mass energies.

Interaction Computations. Computation of interaction energy of two charges involves transformation of effective field strain components from one coordinate system to another, for effecting superposition in a common coordinate system. Let us compute the interaction energy of two electrons located at points O and A with their axes collinear and separated by a distance 'R' along axis OZ (Figure 18.1). Let any space point P be referred to two spherical polar coordinate systems, one (y^i) centered at point O and the other (x^i) centered at point A such that,

$$y^1 \equiv r \ ; \qquad y^2 \equiv \theta \ ; \qquad y^3 \equiv \phi \tag{18.20}$$

and

$$x^1 \equiv \rho \ ; \qquad x^2 \equiv \delta \ ; \qquad x^3 \equiv \phi \tag{18.21}$$

Let the components of any strain tensor referred to coordinate system (y^i) be represented by the symbols S^i_j and those referred to coordinate system (x^i) be represented by the symbols $\in^i_j$. Further, we may designate the field strain components due to the electron located at point O and referred to system (y^i) as $S^i_j(O)$ and those due to the electron located at point A and referred to system (y^i) as $S^i_j(A)$. The superposed or combined strain components at point P may be designated as,

$$S^i_j(C) = S^i_j(O) + S^i_j(A) \tag{18.22}$$

Transformation of the Strain Components. However, before effecting the above superposition we have to first transform the mixed strain tensor components $\in^i_j(A)$ to $S^i_j(A)$ and convert them to the corresponding physical components. The transformation of components $\in^i_j(A)$ in (x^i) coordinate system to $S^i_j(A)$ in (y^i) coordinate system is carried out through the relation,

$$S^i_j(A) = \frac{\partial y^i}{\partial x^\alpha} . \in^\alpha_\beta . \frac{\partial x^\beta}{\partial y^j} \quad \text{(summation over } \alpha \text{ and } \beta) \tag{18.23}$$

For this transformation we need the coordinate transformation relations of the type $y^i = f^i(x^j)$ and $x^i = F^i(y^j)$ between two coordinate systems and the Jacobean matrices of their partial derivatives as under,

$$r^2 = R^2 + \rho^2 + 2 R \rho \cos(\delta) \tag{18.24A}$$

$$\rho^2 = r^2 + R^2 - 2 R r \cos(\theta) \tag{18.24B}$$

$$r \sin(\theta) = \rho \sin(\delta) \quad (18.24C)$$

$$r \cos(\theta) - R = \rho \cos(\delta) \quad (18.24D)$$

$$\tan(\delta) = r.\sin(\theta)/(r.\cos(\theta)-R) \quad (18.24E)$$

Figure 18.1

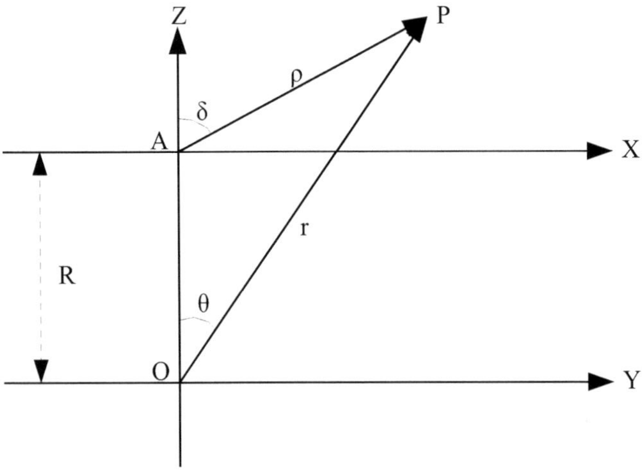

The Jacobean matrices of their partial derivatives obtained from above relations are,

$$\frac{\partial y^1}{\partial x^1}=\frac{\partial r}{\partial \rho}=\frac{r-R.\cos\theta}{\rho} \;;\quad \frac{\partial y^1}{\partial x^2}=\frac{\partial r}{\partial \delta}=-R.\sin\theta \;;\quad \frac{\partial y^1}{\partial x^3}=\frac{\partial r}{\partial \phi}=0$$

$$\frac{\partial y^2}{\partial x^1}=\frac{\partial \theta}{\partial \rho}=\frac{R.\sin\theta}{r.\rho} \;;\quad \frac{\partial y^2}{\partial x^2}=\frac{\partial \theta}{\partial \delta}=\frac{r-R.\cos\theta}{r} \;;\quad \frac{\partial y^2}{\partial x^3}=\frac{\partial \theta}{\partial \phi}=0$$

$$\frac{\partial y^3}{\partial x^1}=\frac{\partial \phi}{\partial \rho}=0 \;;\quad \frac{\partial y^3}{\partial x^2}=\frac{\partial \phi}{\partial \delta}=0 \;;\quad \frac{\partial y^3}{\partial x^3}=\frac{\partial \phi}{\partial \phi}=1 \quad (18.25)$$

And

$$\frac{\partial x^1}{\partial y^1}=\frac{\partial \rho}{\partial r}=\frac{r-R.\cos\theta}{\rho} \;;\quad \frac{\partial x^1}{\partial y^2}=\frac{\partial \rho}{\partial \theta}=\frac{r.R.\sin\theta}{\rho} \;;\quad \frac{\partial x^1}{\partial y^3}=\frac{\partial \rho}{\partial \phi}=0$$

$$\frac{\partial x^2}{\partial y^1}=\frac{\partial \delta}{\partial r}=\frac{-R.\sin\theta}{\rho^2} \;;\quad \frac{\partial x^2}{\partial y^2}=\frac{\partial \delta}{\partial \theta}=\frac{r.(r-R.\cos\theta)}{\rho^2} \;;\quad \frac{\partial x^2}{\partial y^3}=\frac{\partial \delta}{\partial \phi}=0$$

$$\frac{\partial x^3}{\partial y^1}=\frac{\partial \phi}{\partial r}=0 \;;\quad \frac{\partial x^3}{\partial y^2}=\frac{\partial \phi}{\partial \theta}=0 \;;\quad \frac{\partial x^3}{\partial y^3}=\frac{\partial \phi}{\partial \phi}=1 \quad (18.26)$$

From equations (18.19) the effective physical strain components due to the electron located at point O and referred to coordinate system (y^i) are given as,

$$S^{\theta}_{\theta}(O)=S^{\phi}_{\phi}(O)=\frac{1}{\sqrt{2}}.\frac{A_e.e}{r^2}=-S^{r}_{r}(O);$$

$$S^{r}_{\phi}(O)=-\frac{1}{\sqrt{2}}.\frac{A_e.e.\mathrm{Sin}(\theta)}{r^2}=S^{\phi}_{r}(O)$$

$$S^{\phi}_{\theta}(O)=\frac{1}{\sqrt{2}}.\frac{A_e.e.\mathrm{Cos}(\theta)}{r^2}=-S^{\theta}_{\phi}(O); \qquad (18.27)$$

Similarly the effective physical strain components due to the electron located at point A and referred to coordinate system (x^i) are given as,

$$\in^{\delta}_{\delta}(A)=\in^{\phi}_{\phi}(A)=\frac{1}{\sqrt{2}}.\frac{A_e.e}{\rho^2}=-\in^{\rho}_{\rho}(A)$$

$$\in^{\rho}_{\phi}(A)=-\frac{1}{\sqrt{2}}.\frac{A_e.e.\mathrm{Sin}(\delta)}{\rho^2}=\in^{\phi}_{\rho}(A)$$

$$\in^{\phi}_{\delta}(A)=\frac{1}{\sqrt{2}}.\frac{A_e.e.\mathrm{Cos}(\delta)}{\rho^2}=-\in^{\delta}_{\phi}(A) \qquad (18.28)$$

Before using equation (18.23), for transferring the strain components (18.28) from coordinate system (x^i) to the coordinate system (y^i), we have to first convert these physical components to the corresponding mixed tensor $\in^{i}_{j}(A)$ by using the relation (18.10). After thus obtaining $\in^{i}_{j}(A)$, we use equation (18.23) to transform $\in^{i}_{j}(A)$ to $S^{i}_{j}(A)$. The mixed tensor components $S^{i}_{j}(A)$ thus obtained are again converted to the corresponding physical components by using equation (18.10) in coordinate system (y^i) to finally obtain,

$$S^{\theta}_{\theta}(A)=S^{\phi}_{\phi}(A)=\frac{1}{\sqrt{2}}.\frac{A_e.e}{\rho^2}=-S^{r}_{r}(A)$$

$$S^{r}_{\phi}(A)=-\frac{1}{\sqrt{2}}.\frac{A_e.e.\mathrm{Sin}(\theta)}{\rho^2}=S^{\phi}_{r}(A)$$

$$S^{\phi}_{\theta}(A)=\frac{1}{\sqrt{2}}.\frac{A_e.e.\mathrm{Cos}(\theta)}{\rho^2}=-S^{\theta}_{\phi}(A) \qquad (18.29)$$

Interaction Energy Density. At any point P, the strain field energy density due to the electrons located at points O & A is obtained from equation (18.12) as,

$$W_O = \frac{1}{2.\varepsilon_0}.\sum |S^{y^i}_{y^j}(O)|^2 \ ; \ W_A = \frac{1}{2.\varepsilon_0}.\sum |S^{y^i}_{y^j}(A)|^2 \tag{18.30}$$

However, due to the superposition effect, the energy density of the combined field of both electrons located at points O and A is given by,

$$W_C = \frac{1}{2.\varepsilon_0}.\sum |S^{y^i}_{y^j}(O)+S^{y^i}_{y^j}(A)|^2 \tag{18.31}$$

Therefore the interaction energy density W_{int} will be given by,

$$W_{int} = W_C - W_O - W_A \ = \frac{1}{\varepsilon_0}.\sum \left[S^{y^i}_{y^j}(O).S^{y^i}_{y^j}(A) \right] \tag{18.32}$$

This interaction energy density can now be computed from summation of the product of corresponding physical strain components given by equations (18.27) and (18.29), which after simplification is found to be,

$$W_{int} = \frac{(A_e.e)^2}{2\varepsilon_0}.\left[\frac{5}{r^2.\rho^2}\right] \tag{18.33}$$

The total interaction energy E_{int} can now be obtained from volume integral of W_{int} over the entire field of the two interacting charges.

On the other hand if we compute the interaction between a positron and an electron located at points O and A respectively, all the effective strain components given by equations (18.28) and (18.29) will be of the opposite sign, thereby attaching a negative sign to W_{int} of equation (18.33). That is, the interaction energy E_{int} of two opposite charges will be negative. For $R > 0$, the magnitude of E_{int} will be less than twice the field strain energy E_f of one charge and residual field strain energy of the combined field E_C will be greater than zero. Ordinary bulk matter is considered electrically neutral. However, since all negative charges are not completely superposed over positive charges, residual field strain energy E_C of the bulk matter will be finite and greater than zero. This residual field energy might explain the origin of gravitational phenomenon in bulk matter.

Total Interaction Energy. Corresponding to the field interaction energy density W_{int} given by equation (18.33), the total interaction energy E_{int} can be computed by taking volume integral of W_{int} over the entire common field of the interacting electrons as,

$$E_{int} = \int_{\phi=0}^{2\pi} \left[\int_{\theta=0}^{\pi} \int_{r=0}^{\infty} W_{int}.r^2.\sin\theta.dr.d\theta \right] d\phi \tag{18.34}$$

Introducing a dimensionless parameter y = r/R in equations (18.33), (18.34) and (18.24B) we get,

$$E_{int} = 2\pi . R^3 . \int_{\theta=0}^{\pi} \int_{y=0}^{\infty} W_{int} . y^2 . \sin\theta . dy . d\theta$$

$$= \frac{2\pi . (A_e . e)^2}{2\varepsilon_0 . R} . \int_{\theta=0}^{\pi} \int_{y=0}^{\infty} \left[\frac{5 . \sin\theta}{(y^2 + 1 - 2y . \cos\theta)} \right] . dy . d\theta$$

$$= \frac{5\pi . (A_e . e)^2}{\varepsilon_0 . R} . \int_{\theta=0}^{\pi} \left[\tan^{-1}\left(\frac{y - \cos\theta}{\sin\theta} \right) \right]_0^{\infty} d\theta$$

$$= \frac{5\pi . (A_e . e)^2}{\varepsilon_0 . R} . \int_{\theta=0}^{\pi} (\pi - \theta) d\theta = \frac{5\pi^3 . A_e^2 . e^2}{2\varepsilon_0 . R} \qquad (18.35)$$

This verifies the Coulomb interaction law between two electrons as also between two positrons. Similarly, the interaction energy between an electron and a positron can be shown to be given by equation (18.35) but with a negative sign. Mutual force between two charges (electrons or positrons) is given by the negative derivative of equation (18.35) with respect to relative separation R between them.

18.4 Salient Parameters of the Electron

Core Size and Oscillation Frequency. Comparing the interaction energy given by equation (18.35) with the Coulomb interaction energy or the so-called Coulomb potential energy of $e^2/(4\pi\varepsilon_0 R)$ we can compute the dimensionless constant factor 'A_e', which works out to about (1/31.21). Substituting this value of A_e in the total energy relation (18.18), with known mass of the electron $\mathbf{m}_e = 9.109\times10^{-31}$ kg, we obtain the value of κ to be equal to 1.73767×10^{15} m^{-1}. From the relation $x=\kappa.r_c=b_1$ for the core boundary, the electron (positron) core radius is found to be $r_c=1.61\times10^{-15}$ m. Beyond this core boundary of 1.61 fm radius, the strain wave field of the electron extends to infinity and accommodates about 35 percent of its total mass energy. The characteristic frequency of oscillations of the electron/positron core as well as their wave fields, is given by $\nu_e=\kappa.c/2\pi = 8.291\times10^{22}$ Hz. This frequency plays a unique role in all charged particle interactions. The electron/positron type strain bubbles will be able to interact only with those stable/unstable strain bubbles, whose characteristic oscillation frequency matches with ν_e. That means the nucleons and all other charged particles must also be oscillating with this characteristic frequency ν_e . However, since we have not accounted for the effects of spin axis reversal on the field interaction energy or the Coulomb interaction, we need to regard the above mentioned parameters of κ and r_c as tentative at this stage. These parameters may be refined after accounting for the contribution of spin interaction in the overall Coulomb interaction.

Moment of Inertia of the Electron Core. As pointed out above, about 5.88×10^{-31} kg mass of electron is contained in the core region and about 3.229×10^{-31} kg of its mass is spread out in the radial wave field. As an obvious extension of the notion of mass-energy equivalence, we can associate the property of inertia to the total strain energy content as well as the strain energy density in any strain bubble. The inertial property of electron field energy density may have very important consequences. To begin with, we can easily calculate the moment of inertia I_{ec} of the electron core about the Z-axis, by using the mass density (W_c/c^2) from equation (18.17), which works out to be equal to 5.98×10^{-61} kg m^2. However, since the electron mass spread out in its field is not rigidly bound to the core, the notion of moment of inertia is strictly not valid for the radial strain wave field.

Mechanical Spin & Magnetic moment of Electron. The electron core with its finite mass and moment of inertia, can be easily set in translational or rotational motion through various interactions. However, due to the inertial property, the strain field energy will tend to lag behind the moving core. This inertial lag of the 'field' is generally visualized for the translational motion as a consequence of the finite velocity 'c' of the spread of 'electromagnetic field' and gives rise to the well known magnetic field around charged particles in motion. But in the case of rotational motion of the electron core induced by certain interactions, the surrounding radial strain wave field will tend to lag behind due to inertia or equivalently the finite velocity 'c' of phase waves. The rotational motion of the electron core about its axis may be termed as 'mechanical spin' of the electron. Angular lag of the rotating strain wave field, associated with the mechanical spin of the electron, may give rise to the axial magnetic field and the already familiar magnetic moment of the electron.

19

The Photon Wave Packet and Neutrinos

19.1 Equilibrium Equations in Cartesian Coordinates

Introduction. The photon is an elementary particle associated with electromagnetic phenomena. It is the carrier of electromagnetic radiation of all wavelengths, that originate from the atomic or molecular electron transitions. The photon displays both wave and particle properties. As a wave, a single photon is distributed over space and shows wave-like phenomena, such as refraction by a lens. As a particle, it can interact with matter by transferring an amount of energy $E= h.c/\lambda$, where h is Planck's constant, c is the speed of light, and λ is its wavelength. The momentum associated with the photon is given by the relation $p = E/c = h/\lambda$. Although the photon has been the subject of most extensive studies, we still don't know much about its internal structure.

In conventional Cartesian coordinate system (x,y,z), with physical components of the displacement vector **U** given by u^x, u^y and u^z, the equilibrium equations of elasticity are given by a set of three second order partial differential equations as,

$$\partial^2 u^x/\partial x^2 + \partial^2 u^x/\partial y^2 + \partial^2 u^x/\partial z^2 = (1/c^2)\, \partial^2 u^x/\partial t^2 \qquad (19.1A)$$

$$\partial^2 u^y/\partial x^2 + \partial^2 u^y/\partial y^2 + \partial^2 u^y/\partial z^2 = (1/c^2)\, \partial^2 u^y/\partial t^2 \qquad (19.1B)$$

$$\partial^2 u^z/\partial x^2 + \partial^2 u^z/\partial y^2 + \partial^2 u^z/\partial z^2 = (1/c^2)\, \partial^2 u^z/\partial t^2 \qquad (19.1C)$$

These three equations may be grouped into one, involving vector **U** as,

$$\partial^2 \mathbf{U}/\partial x^2 + \partial^2 \mathbf{U}/\partial y^2 + \partial^2 \mathbf{U}/\partial z^2 = \nabla^2 \mathbf{U} = (1/c^2)\, \partial^2 \mathbf{U}/\partial t^2 \qquad (19.1)$$

This is the standard vector wave equation. This wave equation will represent transverse wave propagation when the divergence of **U** is also zero.

$$\nabla.\mathbf{U} = 0 \qquad (19.2)$$

19.2 The Electromagnetic Field

The transverse electromagnetic waves in free space, characterized by zero divergence, are represented by the following standard wave equations,

$$\nabla^2 \mathbf{E} = (1/c^2)\, \partial^2 \mathbf{E}/\partial t^2 \qquad (19.3)$$

$$\nabla^2 \mathbf{B} = (1/c^2)\, \partial^2 \mathbf{B}/\partial t^2 \qquad (19.4)$$

where the electric and magnetic field vectors **E** and **B** satisfy the Maxwell's field equations,

$$\nabla\times\mathbf{E} = -\, \partial \mathbf{B}/\partial t \qquad (19.5)$$

$$\nabla\times\mathbf{B} = (1/c^2).\ \partial\mathbf{E}/\partial t \qquad (19.6)$$

Here, the electric and magnetic field vectors **E** and **B** can be expressed in terms of displacement vector **U** as already discussed in chapter 16.

$$\mathbf{E} = -\ (1/\varepsilon_0).(1/c).\partial\mathbf{U}/\partial t \qquad (19.7)$$

$$\mathbf{B} = (1/c).(1/\varepsilon_0).\ (\nabla\times\mathbf{U}) \qquad (19.8)$$

Through this correlation, in conjunction with equations (19.1) to (19.4), the Maxwell's electromagnetic field equations in free space are identically satisfied by the displacement vector **U** in the elastic space continuum. In essence, that means the electric field vector **E** represents the 'temporal stress' field in the Elastic Space Continuum and is always a function of space and time coordinates. On the other hand, the magnetic field vector **B** represents (1/c) times the 'torsional stress' in the Elastic Space Continuum, and is also a function of space and time coordinates.

Motion Induced Electromagnetic Fields. Let us consider uniform motion of an electron along + X axis, at velocity v. Due to finite velocity c of the phase waves ψ, the complete wave field of the electron will get 'deformed'. This field deformation may be considered through the concept of retarded time and retarded position vector leading to motion induced change in amplitude and direction of the electrostatic field or phase waves at any particular point. Since the kinetic energy of the moving particle will be stored in its deformed field, most of the field strain components are expected to be increased. Let us say that the original field displacement vector **U**, is deformed to **U′** through motion induced field changes. Then **U′** - **U** = **A** may be defined as motion induced displacement field, which will vanish when the particle velocity becomes zero. The motion induced electric and magnetic fields of the moving particle can now be derived from the time derivative and curl of this induced displacement field by using equations (19.7) and (19.8) as,

$$\mathbf{E} = -(1/\varepsilon_0 c).\partial(\mathbf{U}'-\mathbf{U})/\partial t = -(1/\varepsilon_0 c).\partial\mathbf{A}/\partial t \qquad (19.9)$$

and

$$\mathbf{B} = (\mu_0 c).[\nabla\times\mathbf{A}]\ . \qquad (19.10)$$

Under certain conditions of motion and mutual interactions, some part of the motion induced fields could be dissociated from the moving particle, whereas the bound field given by **U** can never be dissociated, unless the particle itself gets annihilated. Normally, the induced fields are an integral part of the moving particle system and it is a matter of interpretation whether the particle motion controls the induced fields or the induced fields govern the particle motion. However we may safely imply that any change in the induced fields will affect the particle motion and any change in particle motion will affect the induced fields. In that sense the motion induced field may be correlated with the de-Broglie waves associated with moving particles.

19.3 The Photon

Induced Field Emission. During certain interactions, if a part of the motion induced fields (**E** and **B** fields) accompanying the electron tend to get separated from the electron system, the separated part of the field must independently satisfy Maxwell's field equations, the vector wave equation, the boundary conditions and also satisfy the overall energy and momentum conservation. For the separated or released field designated by $\mathbf{E}_p$ and $\mathbf{B}_p$, the field conditions will require that $|\mathbf{E}_p| \propto |\mathbf{B}_p|$ and that their strength & spatial spread will be governed by their time rates of change. This implies that the angular frequency ω=kc, of the separated or released field variation, will govern the intensity as well as the spatial spread of $\mathbf{E}_p$ & $\mathbf{B}_p$. The released field wave packet, with specific spatial spread of $\mathbf{E}_p$ & $\mathbf{B}_p$ fields, is the familiar photon wave packet with finite energy content which is proportional to ω.

Plane Wave Solution. Since, the electric and magnetic field vectors **E** & **B** associated with a photon wave packet are in fact derivative functions of the displacement vector **U** the necessary boundary conditions will have to be satisfied by **U**. That is, the displacement vector components u^x, u^y and u^z must vanish at the boundary surface of the photon wave packet and must remain finite and continuous within the packet. The strain energy content within the wave packet must be finite and constant or invariant with respect to time. Most often the electromagnetic waves are visualized as plane waves. Let us consider such a plane wave of angular frequency ω=kc, propagating along +X axis at constant speed c. In terms of displacement vector component u^y, (with $u^x = u^z = 0$) a typical plane wave solution of equilibrium equations (19.1) and (19.2), can be represented as,

$$u^y = A\, e^{i(kx-\omega t)} = A\, e^{i(kx-kct)} = A\, e^{ik(x-ct)} \qquad (19.11)$$

However, this plane wave solution does not satisfy most of the boundary conditions, in that neither the displacement component u^y vanish even at infinity, nor the total strain energy content remain finite for any finite A.

Finite Wavelet Solution. In order to ensure that the displacement vector components u^x, u^y, u^z vanish at the bounding surface of a photon wave packet with finite energy content, we need to modify the plane wave solution (19.11). Let us consider such a modified form as,

$$u^y = A.f(x,y,z,t).e^{ik(x-ct)} \qquad (19,12)$$

where, f(x,y,z,t) is an amplitude decaying function that tends to zero for large values of coordinate parameters, |x|, |y|, |z| and |t|. However, two problems arise with the proposed modification. Firstly the displacement vector components u^y of equation (19.12) no longer satisfy the null divergence condition. Secondly the modified solution must also satisfy the equilibrium equations (19.1). Keeping these in mind, a tentative modified solution of equations (19.1) and (19.2), that meets the required boundary conditions for a photon wave packet, is given below.

$$u^y = A.e.k.[e^{(-k.(1+i).|y|)} . e^{(-k.(1-i).|z|)} . e^{(-k.(1+i).|x-ct|)}] \qquad (19.13)$$

$$u^x = -A.e.k.[e^{(-k.(1+i).|y|)} . e^{(-k.(1-i).|z|)} . e^{(-k.(1+i).|x-ct|)}] \qquad (19.14)$$

$$u^z = 0 \qquad (19.15)$$

Here, the real and imaginary parts of the above complex solution are independently valid solutions of the vector wave equation or the equilibrium equations of elasticity in the space continuum.

It is important to note that apart from the dimensionless constant of integration 'A' in equations (19.13) and (19.14), two dimensional constants, 'e' and k have a special significance in this solution. Whereas 'e', representing the magnitude of electron charge in Coulombs, enables the use of elastic constant $1/\varepsilon_0$ in the standard electrical units, the use of wave number k as an integration constant has double significance. Firstly it ensures the dimensional balance of equations (19.13) and (19.14). Secondly, it represents a common feature of all solutions of equilibrium equations of elasticity in the isotropic elastic space continuum that the maximum amplitude of displacement is *somehow* proportional to the rate of build up or the frequency of such displacement build up. This feature is also responsible for ensuring that the total strain energy content in the photon wave packet is proportional to the frequency of its strain wave oscillations. From equations (19.13) and (19.14), we can compute all the elements of the strain tensor S^i_j. With various strain components given by, $u^x_y = \partial u^x/\partial y$ and $u^x_t = (1/c).\partial u^x/\partial t$ etc. the energy density in the photon wave packet, is:

$$W_{ph} = (1/2\varepsilon_0).[|u^x_x|^2 + |u^x_y|^2 + |u^x_z|^2 + |u^x_t|^2 + |u^y_x|^2 + |u^y_y|^2 + |u^y_z|^2 + |u^y_t|^2]$$

$$= (1/\varepsilon_0).8.A^2.e^2.k^4.e^{(-2k.(|x| + |y| + |z|))} \qquad (19.16)$$

Origin of Planck's constant h. Finally, the total energy content in the wave packet is computed from the volume integral of W_{ph} taken over the entire region of spatial extension of photon. With $\omega = k.c$ this total energy works out to:

$$\mathbf{E_{ph}} = 8\times A^2.e^2.k/\varepsilon_0 = (8/\varepsilon_0 c).A^2.e^2.\omega = \mathbf{(h/2\pi).\omega} = \boldsymbol{\hbar.\omega} \qquad (19.17)$$

Where the factor involving constant terms ε_0, c, A and e, is equated with an important universal constant **h/2π**. The non dimensional constant A can now be evaluated from this relation and it works out to be approximately equal to 1.167 . Origin of Plank's constant **h** is thus linked with the computation of total strain energy in the photon wave packet. Although such a wave packet is emitted from the spatially extended induced field of the electron under certain characteristic interaction conditions, this emission cannot be instantaneous due to finite recoil forces exerted by finite strength fields. It can thus be seen from the spatial extension of equations (19.13) and (19.14), that the photon wave packet is essentially just a small sinusoidal pulse of displacement vector field $\mathbf{U}_p$ with exponentially decaying amplitude. In other words, the photon may be viewed as a sinusoidal pulse of electromagnetic field $\mathbf{E}_p$ & $\mathbf{B}_p$ with exponentially decaying amplitude and 'significant' spatial extension of just about one wavelength in all directions.

The Photon Interaction. Here, it may be appropriate to point out that just like computation of Coulomb Interaction, mutual interaction of two or more photons separated by distance 'd' along any Cartesian coordinate axis, can be easily computed. This is done by superposition of the strain tensor components of two interacting photons separated by distance d along any coordinate axis and referred to a common Cartesian coordinate system. The strain energy of the superposed or combined field can then be easily computed. The computation results show that the interaction energy for two photons of same frequency depends on functions of the type $2\hbar.\omega.e^{(-k.d)}.\cos(k.d)$. That is, any two photons of same frequency ω, will tend to get mutually coupled at certain optimum separation of the order of 'odd number of half wavelengths'. Their interaction energy will change from negative to positive if their separation along any coordinate axis is changed by about one half wave length, resulting in their mutual repulsion. This may account for the conventional interference and dispersion effects encountered in a stream of photons of the same frequency.

19.4 The Neutrinos

Types of Neutrinos. Neutrinos are elementary particles without any electric charge and uncertain rest mass. Traveling close to the speed of light, neutrinos can pass through ordinary matter almost undisturbed and hence, are extremely difficult to detect. Neutrinos are created as a result of certain types of nuclear reactions or radioactive decay or when cosmic rays hit atoms. There are three types of neutrinos: electron neutrinos, muon neutrinos and tau neutrinos along with their anti neutrinos. Electron neutrinos or anti neutrinos are generated whenever neutrons change into protons or vice versa during beta decay. Experimental results show that all observed neutrinos have left-handed helicities (spins anti parallel to momenta), and all anti neutrinos have right-handed helicities.

Distinction between Photons and Neutrinos. Whereas the photons are essentially finite packets of electromagnetic waves, neutrinos are not electromagnetic in nature. Neutrinos are a sort of spinning wave packets. Photons are created from within the wave fields during electromagnetic interactions of charged particles. The neutrinos, on the other hand, are created from within the core regions of various strain bubbles. During the electromagnetic interaction of two charge particles, the conservation of total energy, momentum and angular momentum within the system serves as the main constraint that governs the process of photon creation. Similarly, the process of neutrino creation from within the core region of the associated strain bubbles, will be governed by the conservation of energy, momentum and angular momentum apart from the constraint of equilibrium equations.

Conditions for Emission or Absorption of Neutrinos. Just as the emission and absorption of photons is governed by the energy, momentum and angular momentum transitions of the associated charge particles, the emission and absorption of neutrinos is also governed by the energy, momentum and angular momentum transitions constraints of the associated strain bubble cores. In general, the emission or absorption of a neutrino from within the core region of a strain bubble will occur under following conditions:

- During the transition of a strain bubble core from one energy state to another permissible energy state.
- During the breakup or formation of composite strain bubble cores, an appropriate neutrino will either be released or absorbed to ensure the conservation of energy, momentum and angular momentum. However, during such a transition process, all component strain bubbles will have to satisfy the equilibrium equations as well as the associated boundary conditions.
- During the breakdown or transition of a high energy unstable strain bubble core an appropriate neutrino will be created in addition to other stable or unstable particles. Creation of such a neutrino will be necessary to account for the difference in conserved parameters of the initial and final particles.

Spiraling Wave Packet Representation of Neutrinos. One most prominent spiraling wave packet that appears to represent a neutrino is obtained from cylindrically symmetric solution of equilibrium equations. In cylindrical coordinates, with physical components u^ρ, u^ϕ, u^z of displacement vector **U**, the equilibrium equations (19.1) take on the form of three partial differential equations as follows:

$$\frac{\partial^2 u^\rho}{\partial \rho^2}+\frac{1}{\rho}\cdot\frac{\partial u^\rho}{\partial \rho}-\frac{u^\rho}{\rho^2}+\frac{1}{\rho^2}\cdot\frac{\partial^2 u^\rho}{\partial \phi^2}+\frac{\partial^2 u^\rho}{\partial z^2}-\frac{2}{\rho^2}\cdot\frac{\partial u^\phi}{\partial \phi}=\frac{1}{c^2}\cdot\frac{\partial^2 u^\rho}{\partial t^2}$$

$$\frac{\partial^2 u^\phi}{\partial \rho^2}+\frac{1}{\rho}\cdot\frac{\partial u^\phi}{\partial \rho}-\frac{u^\phi}{\rho^2}+\frac{1}{\rho^2}\cdot\frac{\partial^2 u^\phi}{\partial \phi^2}+\frac{\partial^2 u^\phi}{\partial z^2}+\frac{2}{\rho^2}\cdot\frac{\partial u^\rho}{\partial \phi}=\frac{1}{c^2}\cdot\frac{\partial^2 u^\phi}{\partial t^2}$$

$$\frac{\partial^2 u^z}{\partial \rho^2}+\frac{1}{\rho}\cdot\frac{\partial u^z}{\partial \rho}+\frac{1}{\rho^2}\cdot\frac{\partial^2 u^z}{\partial \phi^2}+\frac{\partial^2 u^z}{\partial z^2}=\frac{1}{c^2}\cdot\frac{\partial^2 u^z}{\partial t^2} \tag{19.18}$$

Here, if we set $u^\rho = u^\phi = 0$, then equation (19.18) can be solved for u^z that consists of a strain wave spiraling along the Z-axis.

$$u^z = \pm A_m.e\kappa.\ J_m(x).\ \mathrm{Cos}(m\phi + qz \pm \kappa ct) \tag{19.19}$$

for $m \geq 1$; $x = (\kappa^2 - q^2)^{1/2}\rho$; $-n\,\pi/2 \leq (m\phi + qz \pm \kappa ct) \leq n\,\pi/2$ and $0 \leq x \leq \alpha_1$ with $J_m(\alpha_1) = 0$ and 'n' a +ve integer.

This type of strain bubble is likely to have negligible interaction with other strain bubbles and will represent a neutrino particle. This is a unique solution that represents a spiraling strain wave that is capable of propagating at up to the speed of light c. Even though the total energy content of this neutrino particle is finite and stable, yet its rest mass may be considered uncertain due to high propagation speed. The wave number κ of this neutrino will be the same as that of the originating strain bubbles. The other characteristic parameters m, n and q will govern the spatial spread, propagation speed and the energy content of this particle.

Spinning Wave Packet Representation of Neutrinos. In spherical polar coordinates, with physical components u^r, u^θ, u^ϕ of vector **U**, the equilibrium equations (19.1) take on the form,

$$\frac{\partial^2 u^r}{\partial r^2}+\frac{2}{r}\cdot\frac{\partial u^r}{\partial r}-\frac{2}{r^2}.u^r+\frac{1}{r^2}\left(\frac{\partial^2 u^r}{\partial \theta^2}+\cot\theta.\frac{\partial u^r}{\partial \theta}+\frac{1}{\sin^2\theta}\cdot\frac{\partial^2 u^r}{\partial \phi^2}\right)$$
$$-\frac{2}{r^2}\left(\frac{\partial u^\theta}{\partial \theta}+\cot\theta.\,u^\theta+\frac{1}{\sin\theta}\cdot\frac{\partial u^\phi}{\partial \phi}\right)=\frac{1}{c^2}\cdot\frac{\partial^2 u^r}{\partial t^2}$$

$$\frac{\partial^2 u^\theta}{\partial r^2}+\frac{2}{r}.\frac{\partial u^\theta}{\partial r}-\frac{u^\theta}{r^2.\sin^2\theta}+\frac{1}{r^2}\left(\frac{\partial^2 u^\theta}{\partial \theta^2}+\cot\theta.\frac{\partial u^\theta}{\partial\theta}+\frac{1}{\sin^2\theta}.\frac{\partial^2 u^\theta}{\partial\phi^2}\right)$$
$$+\frac{2}{r^2}\left(\frac{\partial u^r}{\partial\theta}-\frac{\cot\theta}{\sin\theta}.\frac{\partial u^\phi}{\partial\phi}\right) = \frac{1}{c^2}.\frac{\partial^2 u^\theta}{\partial t^2}$$

$$\frac{\partial^2 u^\phi}{\partial r^2}+\frac{2}{r}.\frac{\partial u^\phi}{\partial r}-\frac{u^\phi}{r^2.\sin^2\theta}+\frac{1}{r^2}\left(\frac{\partial^2 u^\phi}{\partial \theta^2}+\cot\theta.\frac{\partial u^\phi}{\partial\theta}+\frac{1}{\sin^2\theta}.\frac{\partial^2 u^\phi}{\partial\phi^2}\right)$$
$$+\frac{2}{r^2.\sin\theta}\left(\frac{\partial u^r}{\partial\phi}+\cot\theta.\frac{\partial u^\theta}{\partial\phi}\right) = \frac{1}{c^2}.\frac{\partial^2 u^\phi}{\partial t^2}$$

... (19.20)

An important class of spinning wave solutions of equilibrium equations (19.20), with special constraint of null u^ϕ and both u^r as well as u^θ being sinusoidal functions of ϕ and t coordinates, may also represent a neutrino particle with finite rest mass. One such special solution consists of a spinning wave core type strain bubble represented by u^r and u^θ as given below,

$$u^r = A_{e1}.e\kappa.G_1(x).\ Sin(\theta)Cos(\theta).Cos(\phi\pm\kappa ct). \quad (19.21A)$$

$$u^\theta = A_{e1}.e\kappa.G_1(x).\ Sin^2(\theta).Cos(\phi\pm\kappa ct). \quad (19.21B)$$

$$u^\phi = 0\ ; \quad (19.21C)$$

where $x = \kappa\, r$; $G_1(x) = (\cos x - \sin x / x)/x$

and $0 \le x \le a_1$ with $J_{1+½}(a_1) = 0$ or $a_1 = 4.4934$

This strain bubble is expected to be inherently stable and after studying its interaction characteristics, it may be identified as a neutrino particle. In contrast to the spiraling wave solution of equation (19.19), this spinning wave particle can acquire high speeds only through high kinetic energy imparted to it in the emission process. It is important to note here that the emission and absorption of neutrinos takes place through core interactions, or the so called strong interactions of the associated strain bubbles.

20

Nucleon Structure and Strong Interaction

20.1 Cylindrically Symmetric Strain Bubbles

Introduction. In general the term nucleon refers to both protons and neutrons which are the main constituents of the atomic nucleus. However, in the present context we shall use the term nucleon for referring to the nucleon core which is common to both protons and neutrons. The unique short range forces that bind nucleons so securely into nuclei, constitute the strongest class of forces, known as strong interactions. The range of strong interactions is understood to be of the order of 1.7×10^{-15} m or 1.7 fm. Unfortunately, nuclear forces are not as well understood as electrical forces and in consequence the theory of nuclear structure is still primitive as compared with the theory of atomic structure. A common experimental technique used to explore the nuclear structure is the scattering of high energy particle beams from the nucleus. Detailed interpretations of the results of scattering experiments can provide us with a deep insight into the internal structural details of the nucleons. However, a proper interpretation of the observations demands prior knowledge of detailed interaction characteristics of the particles involved in such scattering. Obviously, scattering experiments alone can not provide us complete information about the interaction characteristics as well as the structural details of the interacting particles, without making certain bold assumptions. One most bold assumption which has almost been taken for granted for more than half a century now, is the exchange theory of interactions. The second related assumption is that the electron and positron, being point charges, are not capable of taking part in strong interactions.

Our study of the Elastic Space Continuum has shown that all interactions take place through the superposition of strain fields of interacting particles and are not mediated through the exchange of any particle whatsoever. Therefore, for developing a model of nucleon structure and strong interactions, we first abandon the currently accepted exchange theory of interactions. Further, in the model of Electron Structure & Coulomb Interaction, it has already been brought out that the electron and positron cores will interact mutually or with other strain bubble cores only through strong interactions. It is only the electron positron strain wave fields that interact with other such fields through Coulomb interactions. Possibly therefore, the current interpretations of the observations of various nuclear scattering experiments may have to be revised or refined. The proposed model of nucleon structure is based on a cylindrical strain bubble solution of equilibrium equations of elasticity in the Elastic Space Continuum. This strain bubble is stable, finite in size with cylindrical symmetry and oscillates at a frequency that matches with the oscillation frequency of electron/positron cores.

Equilibrium Equations of Elasticity. Our familiar physical space, with characteristic properties of permittivity ε_0 and permeability μ_0, behaves as a perfect isotropic Elastic Space Continuum with elastic constant $1/\varepsilon_0$ and inertial constant μ_0. The equilibrium equations of elasticity written in terms of displacement vector **U** in this Continuum turn out to be identical to the vector wave equation as given below:

$$\partial^2\mathbf{U}/\partial x^2 + \partial^2\mathbf{U}/\partial y^2 + \partial^2\mathbf{U}/\partial z^2 = \nabla^2\mathbf{U} = (1/c^2)\ \partial^2\mathbf{U}/\partial t^2 \quad (20.1)$$

$$g^{11}u^i_{,11} + g^{22}u^i_{,22} + g^{33}u^i_{,33} = g^{jj}u^i_{,jj} = (1/c^2)\ \partial^2 u^i/\partial t^2 \quad (20.2)$$

The electromagnetic field in the so called 'vacuum' is found to be a dynamic stress-strain field in the corresponding Elastic Space Continuum. A closed region of the Elastic Space Continuum with boundary surface Σ, that satisfies the specified boundary conditions and contains a finite amount of energy in its strain field, is called a 'Strain Bubble'. From the nature of general boundary conditions and the equilibrium equations, it turns out that all valid solutions for displacement vector components u^i are functions of space and time coordinates representing various types of strain wave oscillations. That is, all Strain Bubbles contain a constant finite amount of total strain energy and essentially consist of various strain wave oscillations within a specific boundary surface Σ of the Elastic Space Continuum. One of the cylindrically symmetric solutions of the equilibrium equations, represent the nucleon strain bubble consisting of standing strain wave oscillations.

Solutions with Cylindrical Symmetry. Let us consider a cylindrical coordinate system defined by, $y^1 = \rho$, $y^2 = \phi$ and $y^3 = z$, related to conventional Cartesian coordinates x, y, z as, $x = \rho\cos(\phi)$; $y = \rho\sin(\phi)$; $z = z$. The physical components u^ρ, u^ϕ, u^z of displacement vector **U** are related to its contravariant components u^1, u^2, u^3 as $u^\rho = u^1$; $u^\phi = \rho u^2$; $u^z = u^3$. The physical components of spatial strain in this coordinate system are,

$$S^\rho_\rho = \frac{\partial u^\rho}{\partial \rho} \ ; \ S^\rho_\phi = \frac{1}{\rho}.\frac{\partial u^\rho}{\partial \phi} - \frac{u^\phi}{\rho} \ ; \ S^\rho_z = \frac{\partial u^\rho}{\partial z} \quad (20.3A)$$

$$S^\phi_\rho = \frac{\partial u^\phi}{\partial \rho} \ ; \ S^\phi_\phi = \frac{1}{\rho}.\frac{\partial u^\phi}{\partial \phi} + \frac{u^\rho}{\rho} \ ; \ S^\phi_z = \frac{\partial u^\phi}{\partial z} \quad (20.3B)$$

$$S_{\rho}^{z}=\frac{\partial u^{z}}{\partial \rho} \quad ; \quad S_{\phi}^{z}=\frac{1}{\rho}.\frac{\partial u^{z}}{\partial \phi} \quad ; \quad S_{z}^{z}=\frac{\partial u^{z}}{\partial z} \qquad (20.3C)$$

And the corresponding physical components of temporal strain are given by

$$S_{t}^{\rho}=\frac{1}{c}.\frac{\partial u^{\rho}}{\partial t} \quad ; \quad S_{t}^{\phi}=\frac{1}{c}.\frac{\partial u^{\phi}}{\partial t} \quad ; \quad S_{t}^{z}=\frac{1}{c}.\frac{\partial u^{z}}{\partial t} \qquad (20.3D)$$

For obtaining cylindrically symmetric solutions that are independent of ϕ coordinate, the dynamic equilibrium equations of elasticity can be written in cylindrical coordinates, in terms of physical components (u^{ρ}, u^{ϕ}, u^{z}) of displacement vector **U**, as follows:

$$\frac{\partial^2 u^{\rho}}{\partial \rho^2}+\frac{1}{\rho}.\frac{\partial u^{\rho}}{\partial \rho}-\frac{u^{\rho}}{\rho^2}+\frac{\partial^2 u^{\rho}}{\partial z^2}=\frac{1}{c^2}.\frac{\partial^2 u^{\rho}}{\partial t^2} \qquad (20.4A)$$

$$\frac{\partial u^{\phi}}{\partial \rho^2}+\frac{1}{\rho}.\frac{\partial u^{\phi}}{\partial \rho}-\frac{u^{\phi}}{\rho^2}+\frac{\partial^2 u^{\phi}}{\partial z^2}=\frac{1}{c^2}.\frac{\partial^2 u^{\phi}}{\partial t^2} \qquad (20.4B)$$

$$\frac{\partial^2 u^{z}}{\partial \rho^2}+\frac{1}{\rho}.\frac{\partial u^{z}}{\partial \rho}+\frac{\partial^2 u^{z}}{\partial z^2}=\frac{1}{c^2}.\frac{\partial^2 u^{z}}{\partial t^2} \qquad (20.4C)$$

Solutions of the above equations (20.4) can now be obtained as functions of space-time coordinates, that satisfy the essential conditions of vanishing of vector components (u^{ρ}, u^{ϕ}, u^{z}) at the boundary and time invariance of the strain energy density within the boundary.

20.2 The Nucleon Core

Displacement Vector Field. The most important, lowest order, symmetric solution of equilibrium equations (20.4), which will represent the nucleon core, is

$$u^{\rho} = A_n.e\kappa.\ J_1(x).Cos(qz).Cos(\kappa ct) \quad ; \qquad (20.5A)$$

$$u^{\phi} = A_n.e\kappa.\ J_1(x).Cos(qz).Sin(\kappa ct) \quad ; \quad \& \quad u^{z} = 0 \qquad (20.5B)$$

where A_n is a dimensionless number and $x=(\kappa^2 - q^2)^{½}\rho$. Let us substitute $y=qz$. Here, x and y are dimensionless parameters. The boundary surface Σ is given by $-\pi/2 \le qz \le \pi/2$ & $0 \le x \le \alpha_1$ with $J_1(\alpha_1)=0$ or $\alpha_1=3.832$. Here, 'e' is the magnitude of electron charge and κ is the wave number of strain wave oscillations that is determined from the electron structure to be equal to 1.73767×10^{15} m^{-1}. The strain wave oscillation frequency $\nu_e=\kappa.c/2\pi$ is required to be the same for all mutually interacting particles, like the electron, positron, nucleon and mesons.

The Strain Components. Various strain components for the nucleon core defined by equations (20.5) can now be obtained by using equations (20.3) as given below:

$$S_\rho^\rho(n) = A_n.e\kappa\sqrt{\kappa^2-q^2}.J_1'(x).\mathrm{Cos}(qz).\mathrm{Cos}(\kappa ct)$$

$$S_\phi^\rho(n) = -A_n.e\kappa\sqrt{\kappa^2-q^2}.\frac{J_1(x)}{x}.\mathrm{Cos}(qz).\mathrm{Sin}(\kappa ct)$$

$$S_z^\rho(n) = -A_n.e\kappa q.J_1(x).\mathrm{Sin}(qz).\mathrm{Cos}(\kappa ct)$$

$$S_t^\rho(n) = -A_n.e\kappa^2.J_1(x).\mathrm{Cos}(qz).\mathrm{Sin}(\kappa ct)$$

$$S_\rho^\phi(n) = A_n.e\kappa\sqrt{\kappa^2-q^2}.J_1'(x).\mathrm{Cos}(qz).\mathrm{Sin}(\kappa ct)$$

$$S_\phi^\phi(n) = A_n.e\kappa\sqrt{\kappa^2-q^2}.\frac{J_1(x)}{x}.\mathrm{Cos}(qz).\mathrm{Cos}(\kappa ct)$$

$$S_z^\phi(n) = -A_n.e\kappa q.J_1(x).\mathrm{Sin}(qz).\mathrm{Sin}(\kappa ct)$$

$$S_t^\phi(n) = A_n.e\kappa^2.J_1(x).\mathrm{Cos}(qz).\mathrm{Cos}(\kappa ct) \qquad (20.6)$$

where, $J_1'(x) = J_0(x) - \frac{J_1(x)}{x}$

These strain components of the strain bubble representing a nucleon core, are sinusoidal in time t as well as axial distance z.

Strain Energy Content. The strain energy density in the nucleon core is given by the relation,

$$W_n = \frac{1}{2\varepsilon_0}\left[\text{Sum of the squares of all strain components}\right] \qquad (20.7)$$

That is, from equations (20.6) when we take pair wise sum of the squares of all strain components, Sin(κct) and Cos(κct) terms vanish and we get,

$$W_n = \frac{A_n^2 e^2 \kappa^2}{2\varepsilon_0}\left[(\kappa^2-q^2)\left\{(J_1'(x))^2 + \frac{J_1^2(x)}{x^2}\right\}\mathrm{Cos}^2(qz)\right] + \frac{A_n^2 e^2 \kappa^2}{2\varepsilon_0}\left[J_1^2(x)\left(\kappa^2\mathrm{Cos}^2(qz) + q^2\mathrm{Sin}^2(qz)\right)\right]$$

This is a unique feature of the strain bubble representing a nucleon core that the strain energy density within the core region is completely time invariant implying overall stability of the nucleon. The total strain energy content within the core is computed by integrating W_n over the entire volume of the core. If $2z_1$ is the axial length and ρ_1 is the radius of the core such that $z_1 = \pi/2q$ and $\rho_1 = \alpha_1/(\kappa^2 - q^2)^{1/2}$; then the total energy content E_n will be given by,

$$E_n = \int_0^{2\pi}\int_{-z_1}^{z_1}\int_0^{\rho_1} W_n.\rho.d\rho.dz.d\phi = \frac{4\pi}{(\kappa^2 - q^2)}.\int_0^{z_1}\int_0^{\alpha_1} W_n.x.dx.dz$$

After substituting the values of W_n and z_1 in the above integral and further evaluation we get

$$E_n = \frac{\pi^2 A_n^2 e^2 \kappa \alpha_1^2 J_0^2(\alpha_1)}{2\varepsilon_0 (q/\kappa)(1-(q/\kappa)^2)} \tag{20.8A}$$

This total strain energy function will get minimized for $q/k = 1/\sqrt{3}$. On substituting this value q/κ in the above equations (20.5), (20.6) & (20.8A), we finally get

$$E_n = \frac{3\sqrt{3}\,\pi^2 A_n^2 e^2 \kappa \alpha_1^2 J_0^2(\alpha_1)}{4\varepsilon_0} \quad \text{Joule} \tag{20.8B}$$

Substituting the values of various parameters in the above equation and converting energy units from Joule to MeV, we get $E_n = 960.1836\, A_n^2$ MeV. Assuming this value of total strain energy contained in the nucleon core is almost equal to the known mass energy content of the neutron (**m_n**=939.576 MeV), the dimensionless constant A_n is found to be equal to 0.9892 . With this, the complete internal structure of the nucleon core gets determined.

Size of the Nucleon Core. The overall size of the nucleon core can now be established by substituting the values of different parameters as, $q = \kappa/\sqrt{3}$; $\kappa = 1.73767\times10^{15}\ m^{-1}$; $\alpha_1 = 3.832$ in the relations $z_1 = \pi/2q$ & $\rho_1 = \alpha_1/(\kappa^2 - q^2)^{1/2}$. Thus, we find the maximum length of the nucleon core to be $2z_1$ = 3.1314 fm. The maximum radius of the nucleon core is found to be ρ_1 = 2.7 fm. Hence we find that the nucleon core is of the shape of a right circular cylinder of diameter 5.4 fm and length 3.1314 fm.

Intrinsic Spin Concept. It can be easily seen from the phase quadrature of displacement components u^ρ & u^ϕ in equation (20.5), that the resultant displacement vector keeps continuously rotating or 'spinning' with constant angular velocity $\omega = \kappa c$, whereas, its magnitude remains constant or time invariant at any space point. The direction of this 'spin' of the displacement vector is obviously along the Z-axis of the strain bubble and it remains constant with time. This 'intrinsic spin' of the displacement vector **U** in the nucleon core may, at least partly, be identified with the conventional notion of 'spin' in these particles. Another part of the nucleon spin could be associated with the mechanical rotation of the core about Z-axis, for which its moment of inertia works out to be equal to $I_n = 4.6259\times10^{-57}$ kg.m^2. However, as we shall see later, still another part of the nucleon spin and the anomalous magnetic moment could be associated with the orbital motion of the positron and electron strongly coupled within the nucleon core.

20.3 Strong Interactions

Nature of Strong Interactions. If the strain fields of two strain bubbles overlap in a certain region, the combined strain components will be obtained by superposing corresponding components of both the strain bubbles referred to a common coordinate system. When the cores of two or more strain bubbles get partly overlapped, the resulting interaction is the strong interaction encountered among nucleons and many other elementary particles. The interaction energy (E_{int}) of two such interacting strain bubbles is defined as the difference between the total strain energy of the two strain bubbles with superposed strain fields (E_{sup}) and the sum of separate strain field energies of two bubbles (E_1 and E_2), as:

$$E_{int} = E_{sup} - (E_1 + E_2) \qquad (20.9)$$

If $S^i_j(1)$ and $S^i_j(2)$ represent the strain components of strain bubbles 1 and 2, referred to the same common coordinate system, then it can be seen from equations (20.7) that the interaction energy density W_{int} will be given by the sum of products of the corresponding strain components as,

$$W_{int}(1,2) = (1/2\varepsilon_0).\Sigma[\{ S^i_j(1) + S^i_j(2) \}^2 - \{ S^i_j(1) \}^2 - \{ S^i_j(2) \}^2]$$

$$= (1/\varepsilon_0).\ \Sigma[S^i_j(1).\ S^i_j(2)] \quad (i \to 1 \text{ to } 3\ \&\ j \to 1 \text{ to } 4) \qquad (20.10)$$

Axial n-n Strong Interaction. Let us consider a nucleon core centered at the origin 'O' of a cylindrical coordinate system (y^i), defined by $y^1=\rho$, $y^2=\phi$, $y^3=z$. Let the core axis be aligned along Z-axis (Fig 20.1). From equations (20.5), the non-zero displacement components for this nucleon core termed (O) will be given by,

$$u^\rho(O) = A_n.e\kappa.\ J_1(x).\ Cos(y).\ Cos(\kappa ct) \qquad (20.11A)$$

$$u^\phi(O) = A_n.e\kappa.\ J_1(x).\ Cos(y).\ Sin(\kappa ct) \qquad (20.11B)$$

where $x = \sqrt{2/3}\,(\kappa\rho)$; $\quad y = \sqrt{1/3}\,(\kappa z)$

with $0 \le x \le \alpha_1$; $-\pi/2 \le y \le \pi/2$ and $z_1 = \sqrt{3}.\pi/2\kappa$

The corresponding strain components for this nucleon core are,

$$S^{\rho}_{\rho}(O)=\sqrt{2/3}.A_n.e\kappa^2.J_1'(x).Cos(y).Cos(\kappa ct)$$

$$S^{\rho}_{\phi}(O)=-\sqrt{2/3}.A_n.e\kappa^2.\frac{J_1(x)}{x}.Cos(y).Sin(\kappa ct)$$

$$S^{\rho}_{z}(O)=-\sqrt{1/3}.A_n.e\kappa^2.J_1(x).Sin(y).Cos(\kappa ct)$$

$$S^{\rho}_{t}(O)=-A_n.e\kappa^2.J_1(x).Cos(y).Sin(\kappa ct)$$

$$S^{\phi}_{\rho}(O)=\sqrt{2/3}.A_n.e\kappa^2.J_1'(x).Cos(y).Sin(\kappa ct)$$

$$S^{\phi}_{\phi}(O)=\sqrt{2/3}.A_n.e\kappa^2.\frac{J_1(x)}{x}.Cos(y).Cos(\kappa ct)$$

$$S^{\phi}_{z}(O)=-\sqrt{1/3}.A_n.e\kappa^2.J_1(x).Sin(y).Sin(\kappa ct)$$

$$S^{\phi}_{t}(O)=A_n.e\kappa^2.J_1(x).Cos(y).Cos(\kappa ct) \qquad (20.12)$$

Figure 20.1

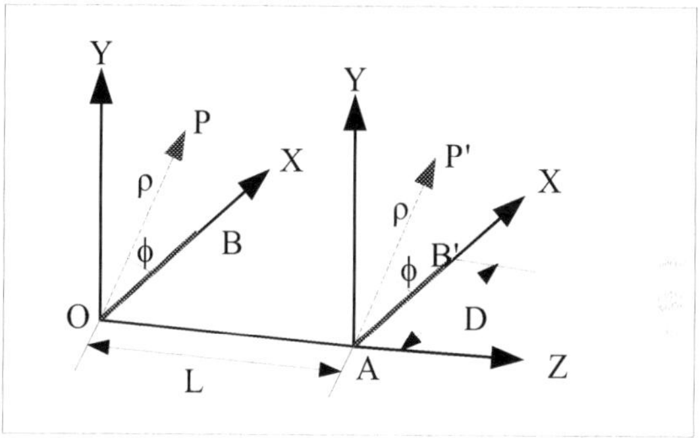

Let us now consider the second nucleon core centered at point 'A' on the z-axis, at distance L from the origin (Fig.20.1), such that its intrinsic spin direction is parallel to that of first core (O). Further, let the displacement vector and strain components of nucleon core (A) be referred to a local cylindrical coordinate system ρ, ϕ, Z such that Z=z-L. Therefore the displacement components for this nucleon core termed (A) will be given by,

$$u^{\rho}(A) = A_n.e\kappa.\ J_1(x).\ Cos(Y).\ Cos(\kappa ct) \qquad (20.13A)$$

$$u^{\phi}(A) = A_n.e\kappa.\ J_1(x).\ Cos(Y).\ Sin(\kappa ct) \qquad (20.13B)$$

where $x=\sqrt{2/3}(\kappa\rho)$ and $Y=\sqrt{1/3}(\kappa Z) = y - \kappa L/\sqrt{3}$ with $0\le x\le\alpha_1$ and $-\pi/2\le Y\le\pi/2$.

Further, let $\delta = L/z_1 = (2\kappa L)/(\pi\sqrt{3})$, so that $Y = y - \delta(\pi/2)$. Corresponding strain components for the nucleon core (A) referred to coordinate system ρ, ϕ, z are:

$$S^{\rho}_{\rho}(A)=\sqrt{2/3}.A_n.e\kappa^2.J_1'(x).Cos(Y).Cos(\kappa ct)$$

$$S^{\rho}_{\phi}(A)=-\sqrt{\frac{2}{3}}.A_n.e\kappa^2.\frac{J_1(x)}{x}.Cos(Y).Sin(\kappa ct)$$

$$S^{\rho}_{z}(A)=-\sqrt{1/3}.A_n.e\kappa^2.J_1(x).Sin(Y).Cos(\kappa ct)$$

$$S^{\rho}_{t}(A)=-A_n.e\kappa^2.J_1(x).Cos(Y).Sin(\kappa ct)$$

$$S^{\phi}_{\rho}(A)=\sqrt{2/3}.A_n.e\kappa^2.J_1'(x).Cos(Y).Sin(\kappa ct)$$

$$S^{\phi}_{\phi}(A)=\sqrt{\frac{2}{3}}A_n.e\kappa^2.\frac{J_1(x)}{x}.Cos(Y).Cos(\kappa ct)$$

$$S^{\phi}_{z}(A)=-\sqrt{1/3}.A_n.e\kappa^2.J_1(x).Sin(Y).Sin(\kappa ct)$$

$$S^{\phi}_{t}(A)=A_n.e\kappa^2.J_1(x).Cos(Y).Cos(\kappa ct) \qquad (20.14)$$

We know from equation (20.10), that the axial interaction energy density W_{ina} in the common overlap region of the two cores, will be given by $(1/\varepsilon_0)$ times the sum of products of the corresponding strain components. Therefore, after pair wise summation of the required products from equations (20.12) and (20.14), we get,

$$W_{ina}(n,n) = \frac{A_n^2.e^2\kappa^4}{\varepsilon_0}.Cos(y)Cos(Y).\left\{\frac{2}{3}.\left((J_1'(x))^2+\frac{J_1^2(x)}{x^2}\right)+J_1^2(x)\right\}$$
$$+\frac{A_n^2.e^2\kappa^4}{\varepsilon_0}.Sin(y)Sin(Y).\left\{\frac{1}{3}.J_1^2(x)\right\}$$

Using the relations:

$$Cos(y)Cos(Y)=(1/2).[Cos(2y-\delta\pi/2)+Cos(\delta\pi/2)]$$

and $$Sin(y)Sin(Y)=(1/2).[Cos(\delta\pi/2)-Cos(2y-\delta\pi/2)]$$

The strain energy density function simplifies to,

$$W_{ina}(n,n) = \frac{A_n^2.e^2\kappa^4}{3\varepsilon_0}.\left[Cos(\delta\pi/2).\left\{\left((J_1'(x))^2+\frac{J_1^2(x)}{x^2}\right)+2J_1^2(x)\right\}\right]$$
$$+\frac{A_n^2.e^2\kappa^4}{3\varepsilon_0}.\left[Cos(2y-\delta\pi/2).\left\{\left((J_1'(x))^2+\frac{J_1^2(x)}{x^2}\right)+J_1^2(x)\right\}\right] \qquad (20.15)$$

Figure 20.2a

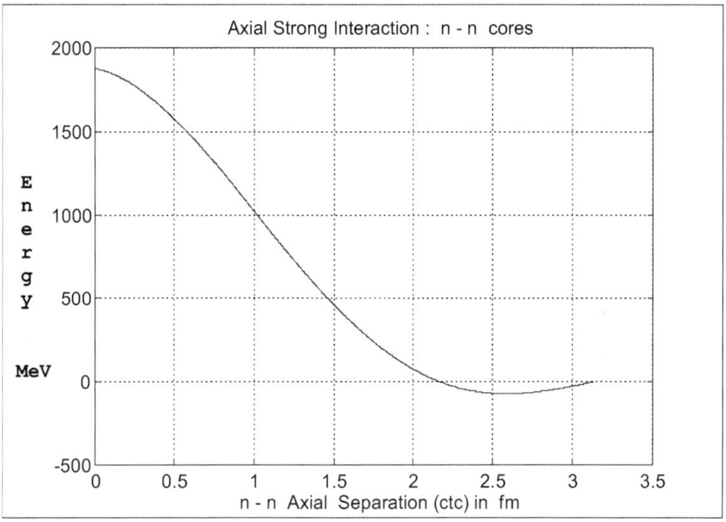

Figure 20.2b

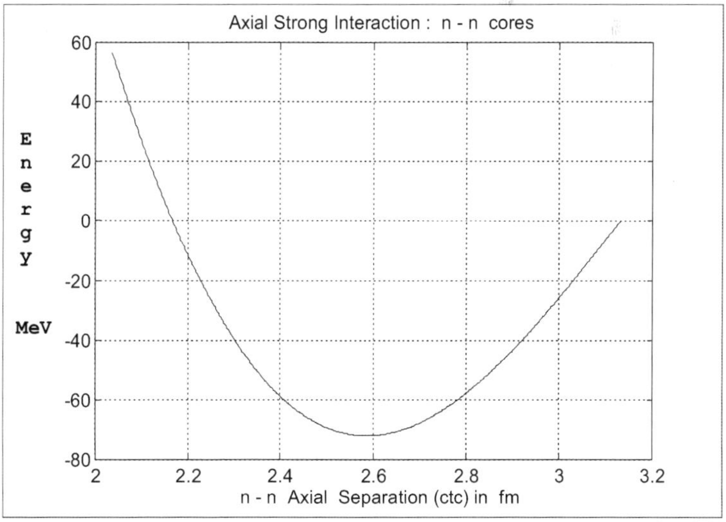

The total axial interaction energy of two nucleon cores is obtained by integrating this energy density over the common overlap region, as

$$E_{ina}(n,n) = \int_0^{2\pi}\int_0^{\rho_1}\int_{L-z_1}^{z_1} W_{ina}.dz.\rho.d\rho.d\phi = 4\pi.\int_0^{\alpha_1}\int_{L/2}^{z_1} W_{ina}.\rho.d\rho.dz$$

$$= \frac{6\pi\sqrt{3}}{\kappa^3}.\int_0^{\alpha_1}\int_{\delta\pi/4}^{\pi/2} W_{ina}.x.dx.dy$$

$$= \frac{\pi\sqrt{3}A_n^2.e^2\kappa.\alpha_1^2.J_0^2(\alpha_1)}{\varepsilon_0}.\left[\frac{3\pi}{2}.\left(1-\frac{\delta}{2}\right).\mathrm{Cos}(\delta\pi/2)+\mathrm{Sin}(\delta\pi/2)\right] \quad (20.16)$$

The total axial interaction energy E_{ina} can be written as a fraction of E_n by using (20.8B), as

$$E_{ina}(n,n) = \left[(2-\delta).\mathrm{Cos}(\delta\pi/2)+(4/3\pi).\mathrm{Sin}(\delta\pi/2)\right].E_n$$

$$= \left[(2-\delta).\mathrm{Cos}(\delta\pi/2)+(4/3\pi).\mathrm{Sin}(\delta\pi/2)\right]\times 939.576 \quad \text{Mev} \quad (20.17)$$

Equation (20.17) gives the axial interaction energy in MeV, as a function of axial separation $\delta=L/z_1$ between the two interacting nucleons. This energy is zero at $\delta=2$ when the nucleons are fully separated and increases to $2E_n$ for $\delta=0$ when the nucleons are fully superposed. For $1.384<\delta<2$ the interaction energy is negative. The exact variation of E_{ina} with δ is shown in figures 20.2. The negative energy part of this axial interaction is an important feature which enables axial bonding between a neutron and a proton. In this p-n coupling, the mean separation between the centers of two cores is about 2.6 fm, varying from about 2.18 fm to 3.13 fm and the frequency of their axial oscillations is about 1.8×10^{21} Hz.

Radial n-n Strong Interaction. Let us again consider a nucleon core centered at the origin 'O' (figure 20.3) of a cylindrical coordinate system (y^i), defined by $y^1=\rho$, $y^2=\phi$, $y^3=z$. Let the core axis be aligned along the Z-axis. The non-zero displacement components for this nucleon core termed (O) are given by equation (20.11) and the corresponding strain components given by equation (20.12). Let us now consider another nucleon core with its axis parallel to the axis of core (O) but separated by distance D. Consider a point B in radial direction $\phi=0$, such that OB=D. Axis of the second core termed core (B) will pass through point B. Further let the center of core (B) be displaced from the plane z=0, along positive Z-axis by distance L. Let the core (B) be referred to a local coordinate system (x^i) defined by $x^1=r$, $x^2=\beta$, $x^3=Z$ such that Z=z-L. Maximum radii of both cores are $\rho_1=r_1$.

Further let $D/\rho_1=\eta$; $L/z_1=\delta$; $x=\sqrt{2/3}.\kappa\rho$; $\chi=\sqrt{2/3}.\kappa r$; $y=\sqrt{1/3}.\kappa z$ $\quad Y=\sqrt{1/3}.\kappa Z=y-\sqrt{1/3}.\kappa L$ $=y-\delta(\pi/2)$.

The coordinates of any point P, located in a common overlap region of two cores, referred to the coordinate systems y^i and x^i, will be inter-related through following transformation relations:

$\rho\,\mathrm{Sin}(\phi)=r\,\mathrm{Sin}(\beta)$; $\qquad \rho\,\mathrm{Cos}(\phi)-D=r\,\mathrm{Cos}(\beta)$;

$\rho^2=r^2+D^2+2rD\,\mathrm{Cos}(\beta)$; $\qquad r^2=\rho^2+D^2-2\rho D\,\mathrm{Cos}(\phi)$;

$$\& \qquad X=\sqrt{x^2+\eta^2\alpha_1^2-2x\,\eta\alpha_1\,\mathrm{Cos}(\phi)} \tag{20.18}$$

Figure 20.3

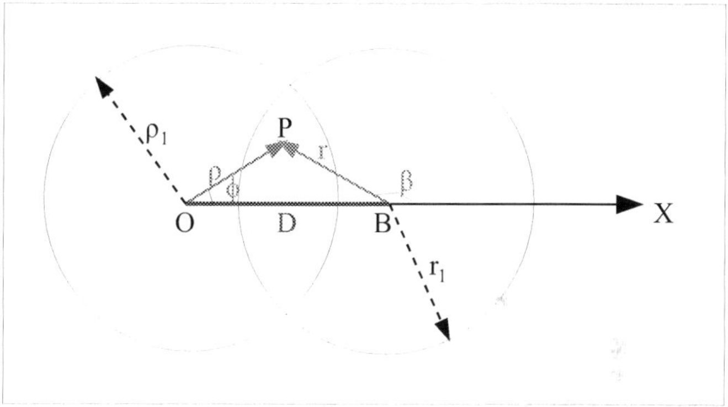

From the above relations, the coordinate transformation Jacobean Matrices of their partial derivatives are obtained as,

$$\frac{\partial y^1}{\partial x^1}=\frac{\partial\rho}{\partial r}=\frac{\rho-D.\cos\phi}{r} \;;\quad \frac{\partial y^1}{\partial x^2}=\frac{\partial\rho}{\partial\beta}=-D.\sin\phi \;;\quad \frac{\partial y^1}{\partial x^3}=\frac{\partial\rho}{\partial Z}=0$$

$$\frac{\partial y^2}{\partial x^1}=\frac{\partial\phi}{\partial r}=\frac{D.\sin\phi}{r.\rho} \;;\quad \frac{\partial y^2}{\partial x^2}=\frac{\partial\phi}{\partial\beta}=\frac{\rho-D.\cos\phi}{\rho} \;;\quad \frac{\partial y^2}{\partial x^3}=\frac{\partial\phi}{\partial Z}=0$$

$$\frac{\partial y^3}{\partial x^1}=\frac{\partial z}{\partial r}=0 \;;\quad \frac{\partial y^3}{\partial x^2}=\frac{\partial z}{\partial\beta}=0 \;;\quad \frac{\partial y^3}{\partial x^3}=\frac{\partial z}{\partial Z}=1 \tag{20.19}$$

And

$$\frac{\partial x^1}{\partial y^1}=\frac{\partial r}{\partial\rho}=\frac{\rho-D.\cos\phi}{r} \;;\quad \frac{\partial x^1}{\partial y^2}=\frac{\partial r}{\partial\phi}=\frac{\rho.D.\sin\phi}{r} \;;\quad \frac{\partial x^1}{\partial y^3}=\frac{\partial r}{\partial z}=0$$

$$\frac{\partial x^2}{\partial y^1}=\frac{\partial \beta}{\partial \rho}=\frac{-D.\sin\phi}{r^2}\ ;\ \frac{\partial x^2}{\partial y^2}=\frac{\partial \beta}{\partial \phi}=\frac{\rho.(\rho-D.\cos\phi)}{r^2}\ ;\ \frac{\partial x^2}{\partial y^3}=\frac{\partial \beta}{\partial z}=0$$

$$\frac{\partial x^3}{\partial y^1}=\frac{\partial Z}{\partial \rho}=0\ ;\ \frac{\partial x^3}{\partial y^2}=\frac{\partial Z}{\partial \phi}=0\ ;\ \frac{\partial x^3}{\partial y^3}=\frac{\partial Z}{\partial z}=1 \qquad (20.20)$$

The displacement components for the second nucleon core termed (B) and referred to coordinate system x^i (r, β, Z) are therefore given by,

$$u^r(B) = A_n.e\kappa.\ J_1(\chi).\ Cos(Y).\ Cos(\kappa ct) \qquad (20.21A)$$

$$u^\beta(B) = A_n.e\kappa.\ J_1(\chi).\ Cos(Y).\ Sin(\kappa ct) \qquad (20.21B)$$

where $\chi=\sqrt{2/3}.\kappa r$; $Y=\sqrt{1/3}(\kappa Z)=y-\delta(\pi/2)$;
with $0\leq\chi\leq\alpha_1$ and $-\pi/2\leq Y\leq\pi/2$.

The physical strain components for the core (B) referred to coordinate system r, β, Z are,

$$\varepsilon_r^r(B)=\sqrt{2/3}.A_n.e\kappa^2.J_1'(\chi).Cos(Y).Cos(\kappa ct)$$

$$\varepsilon_\beta^r(B)=-\sqrt{\frac{2}{3}}.A_n.e\kappa^2.\frac{J_1(\chi)}{\chi}.Cos(Y).Sin(\kappa ct)$$

$$\varepsilon_Z^r(B)=-\sqrt{1/3}.A_n.e\kappa^2.J_1(\chi).Sin(Y).Cos(\kappa ct)$$

$$\varepsilon_t^r(B)=-A_n.e\kappa^2.J_1(\chi).Cos(Y).Sin(\kappa ct)$$

$$\varepsilon_r^\beta(B)=\sqrt{2/3}.A_n.e\kappa^2.J_1'(\chi).Cos(Y).Sin(\kappa ct)$$

$$\varepsilon_\beta^\beta(B)=\sqrt{\frac{2}{3}}A_n.e\kappa^2.\frac{J_1(\chi)}{\chi}.Cos(Y).Cos(\kappa ct)$$

$$\varepsilon_Z^\beta(B)=-\sqrt{1/3}.A_n.e\kappa^2.J_1(\chi).Sin(Y).Sin(\kappa ct)$$

$$\varepsilon_t^\beta(B)=A_n.e\kappa^2.J_1(\chi).Cos(Y).Cos(\kappa ct) \qquad (20.22)$$

Now these strain components have to be transformed to coordinate system y^i(ρ, ϕ, z) centered at O by using the relations,

$$S_j^i(B)=\frac{\partial y^i}{\partial x^\alpha}.\varepsilon_\beta^\alpha(B).\frac{\partial x^\beta}{\partial y^j} \quad \text{(summation over } \alpha \text{ and } \beta) \qquad (20.23)$$

However, before carrying out this transformation we have to first convert the physical strain components $\varepsilon_{x^j}^{x^i}(B)$ given by equations (20.22), to the corresponding strain tensor components $\varepsilon_j^i(B)$ through the relation,

$$\varepsilon_{x^j}^{x^i}(B)\ =\ \sqrt{g_{ii}}.\varepsilon_j^i(B).\sqrt{g^{jj}} \quad \text{(no summation over i or j)} \qquad (20.24)$$

Again after using equation (20.23) the strain tensor components $S^i_j(B)$ will have to be converted back to the physical components $S^{y^i}_{y^j}(B)$ using the above relation.

Finally, the strain components (20.22) due to the core (B), when transformed by using equations (20.19), (20.20), (20.23) and (20.24) to the coordinate system $y^i(\rho, \phi, z)$ centered at O, are obtained as,

$$S^{\rho}_{\rho}(B) = F_1(Y)\left[\left\{\frac{(x-\eta\alpha_1 \mathrm{Cos}(\phi))^2}{\chi^2} J_1'(\chi) + \frac{\eta^2\alpha_1^2 \mathrm{Sin}^2(\phi)}{\chi^3} J_1(\chi)\right\}\mathrm{Cos}(\kappa ct)\right]$$
$$+F_1(Y)\left[\left\{\frac{(x-\eta\alpha_1 \mathrm{Cos}(\phi))\eta\alpha_1 \mathrm{Sin}(\phi)}{\chi^2}\left(\frac{J_1(\chi)}{\chi} - J_1'(\chi)\right)\right\}\mathrm{Sin}(\kappa ct)\right]$$

where $F_1(Y) = \sqrt{\frac{2}{3}} A_n e\kappa^2 \mathrm{Cos}(Y)$

$$S^{\phi}_{\rho}(B) = F_1(Y)\left[\left\{\frac{(x-\eta\alpha_1 \mathrm{Cos}(\phi))^2}{\chi^2} J_1'(\chi) + \frac{\eta^2\alpha_1^2 \mathrm{Sin}^2(\phi)}{\chi^3} J_1(\chi)\right\}\mathrm{Sin}(\kappa ct)\right]$$
$$-F_1(Y)\left[\left\{\frac{(x-\eta\alpha_1 \mathrm{Cos}(\phi))\eta\alpha_1 \mathrm{Sin}(\phi)}{\chi^2}\left(\frac{J_1(\chi)}{\chi} - J_1'(\chi)\right)\right\}\mathrm{Cos}(\kappa ct)\right]$$

$$S^{\rho}_{\phi}(B) = -F_1(Y)\left[\left\{\frac{\eta^2\alpha_1^2 \mathrm{Sin}^2(\phi)}{\chi^2} J_1'(\chi) + \frac{(x-\eta\alpha_1 \mathrm{Cos}(\phi))^2}{\chi^3} J_1(\chi)\right\}\mathrm{Sin}(\kappa ct)\right]$$
$$-F_1(Y)\left[\left\{\frac{(x-\eta\alpha_1 \mathrm{Cos}(\phi))\eta\alpha_1 \mathrm{Sin}(\phi)}{\chi^2}\left(\frac{J_1(\chi)}{\chi} - J_1'(\chi)\right)\right\}\mathrm{Cos}(\kappa ct)\right]$$

$$S^{\phi}_{\phi}(B) = F_1(Y)\left[\left\{\frac{\eta^2\alpha_1^2 \mathrm{Sin}^2(\phi)}{\chi^2} J_1'(\chi) + \frac{(x-\eta\alpha_1 \mathrm{Cos}(\phi))^2}{\chi^3} J_1(\chi)\right\}\mathrm{Cos}(\kappa ct)\right]$$
$$-F_1(Y)\left[\left\{\frac{(x-\eta\alpha_1 \mathrm{Cos}(\phi))\eta\alpha_1 \mathrm{Sin}(\phi)}{\chi^2}\left(\frac{J_1(\chi)}{\chi} - J_1'(\chi)\right)\right\}\mathrm{Sin}(\kappa ct)\right]$$

$$S^{\rho}_{z}(B) = -F_2(Y)\left[\frac{(x-\eta\alpha_1 \mathrm{Cos}(\phi))}{\chi}\mathrm{Cos}(\kappa ct) - \frac{\eta\alpha_1 \mathrm{Sin}(\phi)}{\chi}\mathrm{Sin}(\kappa ct)\right]$$

where $F_2(Y) = \sqrt{\frac{1}{3}} A_n e\kappa^2 J_1(\chi)\mathrm{Sin}(Y)$

$$S_z^{\phi}(B) = -F_2(Y)\left[\frac{(x-\eta\alpha_1 \operatorname{Cos}(\phi))}{\chi}\operatorname{Sin}(\kappa ct) + \frac{\eta\alpha_1 \operatorname{Sin}(\phi)}{\chi}\operatorname{Cos}(\kappa ct)\right]$$

$$S_t^{\rho}(B) = -F_3(Y)\left[\frac{(x-\eta\alpha_1 \operatorname{Cos}(\phi))}{\chi}\operatorname{Sin}(\kappa ct) + \frac{\eta\alpha_1 \operatorname{Sin}(\phi)}{\chi}\operatorname{Cos}(\kappa ct)\right]$$

where $F_3(Y) = A_n e\kappa^2 J_1(\chi)\operatorname{Cos}(Y)$

$$S_t^{\phi}(B) = F_3(Y)\left[\frac{(x-\eta\alpha_1 \operatorname{Cos}(\phi))}{\chi}\operatorname{Cos}(\kappa ct) - \frac{\eta\alpha_1 \operatorname{Sin}(\phi)}{\chi}\operatorname{Sin}(\kappa ct)\right] \qquad (20.25)$$

From equation (20.10), the interaction energy density is given by the sum of products of the corresponding strain components. Therefore, for computing the radial interaction energy of two cores (O) and (B), we may first compute the sum of pairs of products of the corresponding strain components from equations (20.12) and (20.25) as follows:

$$\Sigma_\rho(n,n) = S_\rho^{\rho}(O).S_\rho^{\rho}(B) + S_\rho^{\phi}(O).S_\rho^{\phi}(B)$$

$$= \frac{2}{3}A J_1'(x)\operatorname{Cos}(y)\operatorname{Cos}(Y)\left[\frac{(x-\eta\alpha_1 \operatorname{Cos}(\phi))^2}{\chi^2}J_1'(\chi) + \frac{\eta^2\alpha_1^2 \operatorname{Sin}^2(\phi)}{\chi^3}J_1(\chi)\right]$$

where, $A = A_n^2 e^2 \kappa^4$

$$\Sigma_\phi(n,n) = S_\phi^{\rho}(O).S_\phi^{\rho}(B) + S_\phi^{\phi}(O).S_\phi^{\phi}(B)$$

$$= \frac{2}{3}A\frac{J_1(x)}{x}\operatorname{Cos}(y)\operatorname{Cos}(Y)\left[\frac{\eta^2\alpha_1^2 \operatorname{Sin}^2(\phi)}{\chi^2}J_1'(\chi) + \frac{(x-\eta\alpha_1 \operatorname{Cos}(\phi))^2}{\chi^3}J_1(\chi)\right]$$

$$\Sigma_z(n,n) = S_z^{\rho}(O).S_z^{\rho}(B) + S_z^{\phi}(O).S_z^{\phi}(B)$$

$$= A J_1(x)J_1(\chi)\operatorname{Sin}(y)\operatorname{Sin}(Y)\left[\frac{(x-\eta\alpha_1 \operatorname{Cos}(\phi))}{3\chi}\right]$$

$$\Sigma_t(n,n) = S_t^{\rho}(O).S_t^{\rho}(B) + S_t^{\phi}(O).S_t^{\phi}(B)$$

$$= A J_1(x)J_1(\chi)\operatorname{Cos}(y)\operatorname{Cos}(Y)\left[\frac{(x-\eta\alpha_1 \operatorname{Cos}(\phi))}{\chi}\right]$$

Hence, the interaction energy density in the common overlap region is,

$$W_{inr}(n,n) = (1/\varepsilon_0)[\Sigma_\rho(n,n) + \Sigma_\phi(n,n) + \Sigma_z(n,n) + \Sigma_t(n,n)] \qquad (20.26)$$

This energy density function is completely time invariant and is a function of independent parameters ρ, ϕ, z, δ and η . If the intrinsic spin direction of one of the cores is reversed, then the resulting interaction energy density will no longer be time invariant, leading to overall instability of such interaction. Hence, all meaningful and stable strong interactions among nucleons occur with their intrinsic spin directions parallel. As usual the total interaction energy of the strong interaction considered above, will be obtained by integrating W_{inr} over the entire volume of the common overlapped region, as given below:

$$E_{inr}(n,n) = \iiint W_{inr}(n,n).\rho.d\rho.d\phi.dz \qquad (20.27)$$

The volume integral at equation (20.27) above, with variable limits of integration, can not be evaluated analytically and we must take recourse to numerical integration with the aid of a computer. For any given set of separation parameters δ and η, a unique value of interaction energy will be obtained. If the radial separation parameter η is set to zero, χ will reduce to x and the interaction energy obtained from equation (20.27) will reduce to axial interaction energy given by equation (20.17). If on the other hand δ is set to zero, Y will reduce to y and equation (20.27) will then provide pure radial interaction energy.

Figure 20.4a

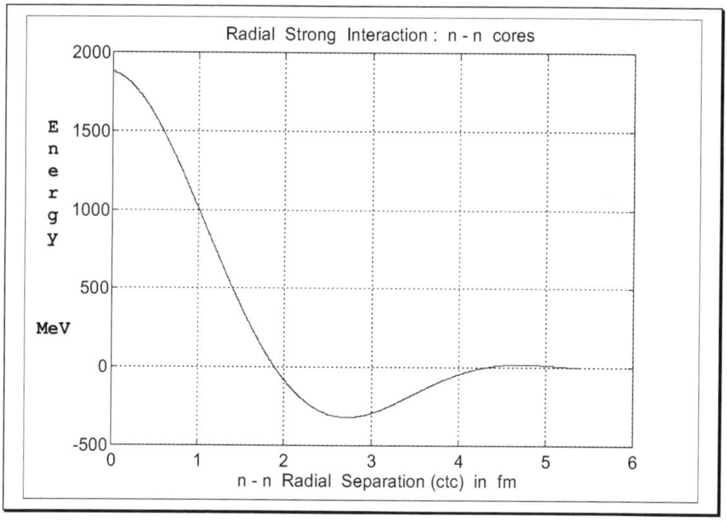

The radial interaction energy between two nucleon cores, thus obtained, is plotted at figure 20.4 for $0 \leq \eta \leq 2$. This is an important plot which shows maximum negative interaction energy between two nucleons at radial separation of one core radius, that is about 2.7 fm. It means that two nucleons are likely to get radially coupled together under suitable conditions and oscillate at a frequency of about 3.6×10^{22} Hz, with mean radial separation of about 2.7 fm. The radial interaction energy is positive for core separation between 0 to 1.9 fm and again from about 4.3 fm to the maximum limit of two core radii. The asymmetry of negative portion of this interaction energy curve with respect to its minima at 2.7 fm, may lead to the rotational motion of the interacting nucleons alongside their above mentioned radial oscillations. The positive 'hump' of this interaction energy curve from 4.3 fm to 5.4 fm separation suggests that the radial coupling between two interacting nucleons may be very difficult to form and once formed may be much more difficult to break in comparison with their axial coupling.

Figure 20.4b

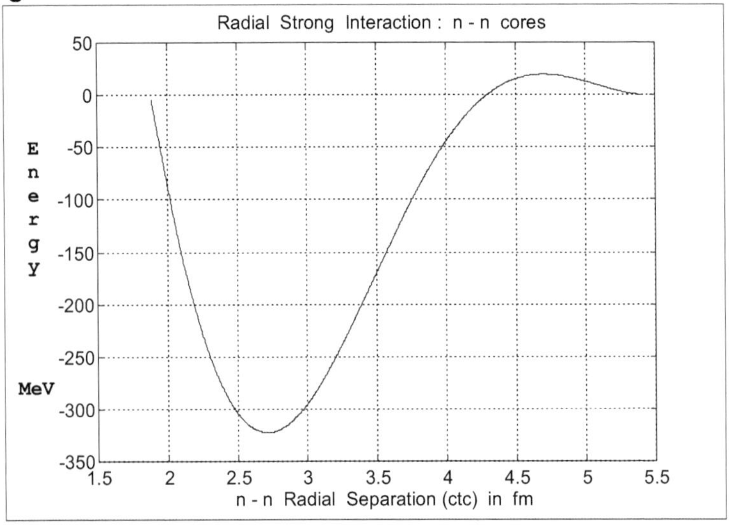

There is one very important point to be noted here regarding the magnitude of negative interaction energy available for strong coupling and the actual 'bond energy' ensuring stable binding between the interacting nucleons. The available interaction energy 'released' by the system, must be actually emitted out of the system for it to become 'bond energy' of the coupled nucleons. In practice, only a small fraction of the negative interaction energy released by the system is actually emitted out of the system and the balance is converted to the kinetic energy of motion of the interacting nucleons. There must be a valid mechanism available for emitting a portion of the released interaction energy out of the system. The interacting nucleons by themselves do not appear to possess any such mechanism. Energy could be emitted out of the system with due conservation of energy and momentum, either as photons or some other elementary particles like mesons, neutrinos etc. But photons could be emitted out from the strain wave field region only through specific relative motion of charge particles - the electron and positron, whereas the neutrinos and mesons could be formed and emitted out from within the core region itself. Therefore, the presence of positron or electron among the nucleons is an essential feature in the coupling and securely binding the nucleons in the nucleus. Hence, we next examine the strong interaction between the positron core and the nucleon core.

20.4 The Proton

The proton is known to be a positively charged nucleon. The electron and positron are stable elementary charge particles, consisting of a spherically symmetric strain wave core, surrounded by a radially propagating phase wave field. As such, it is quite reasonable to expect that the proton may consist of a nucleon core with a positron superposed over it through strong interaction. We may therefore, examine the strong interaction characteristics of the positron core with the nucleon core. We have already examined the strong interaction characteristics of two nucleon cores. Proceeding on the same lines, let us replace the nucleon core centered at the origin O of the cylindrical coordinate system, with a positron core centered at the same point O.

The positron type strain bubbles are obtained as spherically symmetric lowest order solutions of equilibrium equations of elasticity in the Space Continuum. The non-zero displacement vector components of the positron core, referred to a spherical coordinate system (R,θ, ϕ), with origin at O, are given by,

$$u^R = A_e.e\kappa.G_1(X).\ Cos(\kappa ct); \qquad (20.28A)$$

$$u^\phi = A_e.e\kappa.G_1(X).\ Sin(\theta).\ Sin(\kappa ct); \qquad (20.28B)$$

where, $G_1(X) = (\cos X - \sin X / X)/X = - (\pi/2X)^{1/2}.\ J_{3/2}(X)$ and $X = \kappa R$.

The corresponding strain components for the positron core can be properly transformed to a cylindrical coordinate system (ρ, ϕ, z) with origin at point O, by the usual procedure discussed above. These strain components for the positron (e^+) core centered at point O, after proper transformation to the cylindrical coordinate system y^i (ρ, ϕ, z) are:

$$S^\rho_\rho(e^+) = \frac{A_e e\kappa^2}{X^2}\left[\frac{3}{2}x^2 G_1'(X) + 3y^2\frac{G_1(X)}{X}\right].Cos(\kappa ct)$$

$$S^\phi_\rho(e^+) = \frac{A_e e\kappa^2}{X^2}\left[\frac{3}{2}x^2 G_1'(X) + 3y^2\frac{G_1(X)}{X}\right].Sin(\kappa ct)$$

$$S_\phi^\rho(e^+) = -A_e e\kappa^2 \left[\frac{G_1(X)}{X}\right].\mathrm{Sin}(\kappa ct)$$

$$S_\phi^\phi(e^+) = A_e e\kappa^2 \left[\frac{G_1(X)}{X}\right].\mathrm{Cos}(\kappa ct)$$

$$S_z^\rho(e^+) = \frac{3A_e e\kappa^2 xy}{\sqrt{2}.X^2}.\left[G_1'(X) - \frac{G_1(X)}{X}\right].\mathrm{Cos}(\kappa ct)$$

$$S_z^\phi(e^+) = \frac{3A_e e\kappa^2 xy}{\sqrt{2}.X^2}.\left[G_1'(X) - \frac{G_1(X)}{X}\right].\mathrm{Sin}(\kappa ct)$$

$$S_t^\rho(e^+) = -A_e e\kappa^2.\sqrt{\frac{3}{2}}.x.\left[\frac{G_1(X)}{X}\right].\mathrm{Sin}(\kappa ct)$$

$$S_t^\phi(e^+) = A_e e\kappa^2.\sqrt{\frac{3}{2}}.x.\left[\frac{G_1(X)}{X}\right].\mathrm{Cos}(\kappa ct)$$

$$S_\rho^z(e^+) = \frac{3A_e e\kappa^2 xy}{\sqrt{2}.X^2}.\left[G_1'(X) - \frac{G_1(X)}{X}\right].\mathrm{Cos}(\kappa ct)\ ;\ S_\phi^z(e^+) = 0$$

$$S_z^z(e^+) = \frac{A_e e\kappa^2}{X^2}.\left[3y^2 G_1'(X) + \frac{3}{2}x^2\frac{G_1(X)}{X}\right].\mathrm{Cos}(\kappa ct)$$

$$S_t^z(e^+) = -\sqrt{3}A_e e\kappa^2 y.\left[\frac{G_1(X)}{X}\right].\mathrm{Sin}(\kappa ct) \qquad (20.29)$$

Where, $R = \sqrt{(\rho^2+z^2)}$; $X = \kappa R = \kappa.\sqrt{(\rho^2+z^2)} = \sqrt{((3/2)x^2+3y^2)}$
with $x = \sqrt{(2/3)}.\kappa\rho$ and $y = \sqrt{(1/3)}.\kappa z$

From equation (20.10), the interaction energy density in the common overlapped region of the interacting cores is given by $(1/\varepsilon_0)$ times the sum of products of the corresponding strain components. Therefore, for computing the interaction energy of the positron (e^+) centered at O and the nucleon core (n) located at B, with their intrinsic spins parallel and axes separated by distance $D = \eta\rho_1$, we may first compute the sum of pairs of products of the corresponding strain components from equations (20.29) and (20.25) as follows.

$$\Sigma_\rho(e^+,n) = S_\rho^\rho(e^+).S_\rho^\rho(n) + S_\rho^\phi(e^+).S_\rho^\phi(n) = \sqrt{\frac{2}{3}}\frac{A_e A_n e^2\kappa^4 \mathrm{Cos}(Y)}{X^2}.$$

$$\left\{\frac{3x^2}{2}G_1'(X) + 3y^2\frac{G_1(X)}{X}\right\}\left[\frac{(x-\eta\alpha_1\mathrm{Cos}(\phi))^2}{\chi^2}J_1'(\chi) + \frac{\eta^2\alpha_1^2\mathrm{Sin}^2(\phi)}{\chi^3}J_1(\chi)\right]$$

$$\Sigma_\phi(e^+,n) = S^\rho_\phi(e^+).S^\rho_\phi(n)+S^\phi_\phi(e^+).S^\phi_\phi(n) = \sqrt{\frac{2}{3}}A_eA_ne^2\kappa^4\,\text{Cos}(Y).$$

$$\left\{\frac{G_1(X)}{X}\right\}\left[\frac{\left(x-\eta\alpha_1\,\text{Cos}(\phi)\right)^2}{\chi^3}J_1(\chi)+\frac{\eta^2\alpha_1^2\,\text{Sin}^2(\phi)}{\chi^2}J_1'(\chi)\right]$$

$$\Sigma_z(e^+,n) = S^\rho_z(e^+).S^\rho_z(n)+S^\phi_z(e^+).S^\phi_z(n)$$

$$= \sqrt{\frac{3}{2}}A_eA_ne^2\kappa^4.\frac{xy}{X^2}J_1(\chi)\text{Sin}(Y)\frac{\left(x-\eta\alpha_1\,\text{Cos}(\phi)\right)}{\chi}\left\{\frac{G_1(X)}{X}-G_1'(X)\right\}$$

$$\Sigma_t(e^+,n) = S^\rho_t(e^+).S^\rho_t(n)+S^\phi_t(e^+).S^\phi_t(n)$$

$$= \sqrt{\frac{3}{2}}A_eA_ne^2\kappa^4x\frac{G_1(X)}{X}J_1(\chi)\text{Cos}(Y)\frac{\left(x-\eta\alpha_1\,\text{Cos}(\phi)\right)}{\chi}$$

Where, $G_1'(X)=-\left(\frac{\text{Sin}(X)}{X}+\frac{2G_1(X)}{X}\right)$

And $J_1'(\chi)=J_0(\chi)-\frac{J_1(\chi)}{\chi}$

Hence, the interaction energy density in the common overlap region is,

$$W_{int}(e^+,n) = (1/\varepsilon_0)[\Sigma_\rho(e^+,n) + \Sigma_\phi(e^+,n) + \Sigma_z(e^+,n) + \Sigma_t(e^+,n)] \qquad (20.30)$$

Figure 20.5

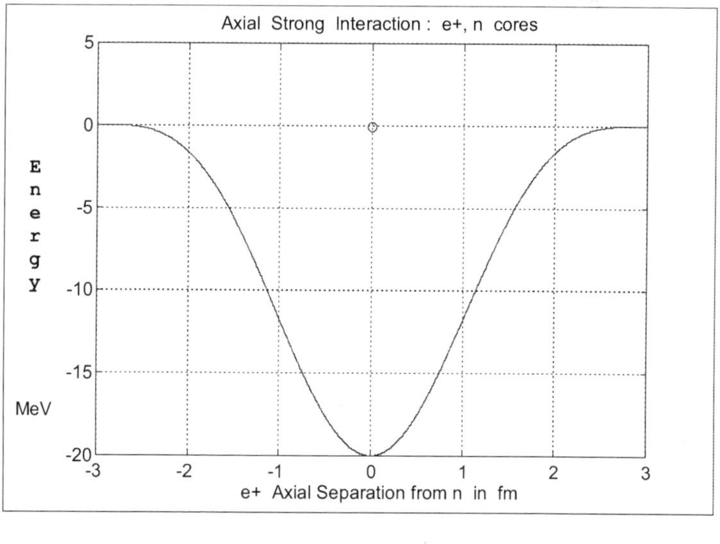

This energy density function is time invariant and is a function of independent parameters ρ, ϕ, z, δ and η. As usual, the total interaction energy of the strong interaction between a nucleon and a positron considered above, will be obtained by integrating $W_{int}(e^+,n)$ over the entire volume of the common overlapped region.

$$E_{int}(e^+,n) = \iiint W_{int}(e^+,n).\rho.d\rho.d\phi.dz \qquad (20.31)$$

The above volume integral, with variable limits of integration for the common overlapped region of two cores, can be evaluated through computerized numerical integration. For any set of separation parameters δ and η, a unique value of interaction energy will be obtained. If the radial separation parameter η is set to zero, axial interaction energy of the e^+,n pair will be obtained from equation (20.31). If δ is set to zero, equation (20.31) will provide pure radial interaction energy.

Figure 20.6

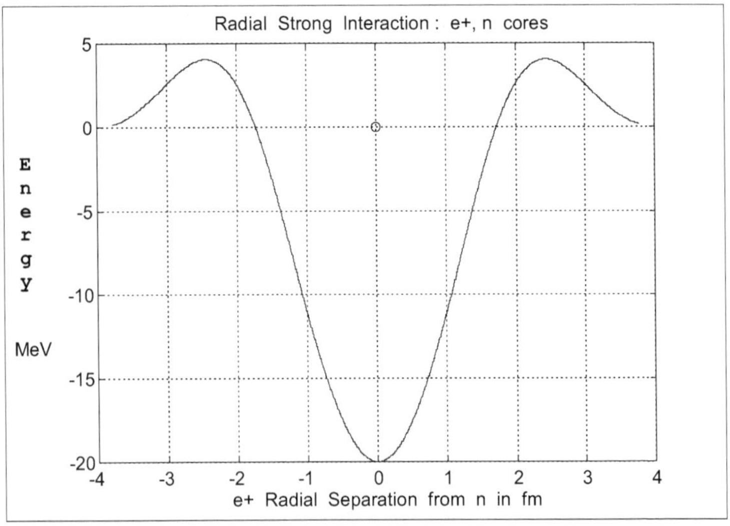

The axial and radial interaction energy of the positron, nucleon cores thus obtained is plotted at figures 20.5 & 20.6 for various separation distances. These are important plots which show maximum negative interaction energy of about 20 MeV between two cores when their centers coincide. The radial interaction energy is positive for core separation between 1.7 to 3.7 fm. We may even compute the whole set of values of interaction energy between the positron and nucleon cores for all possible values of δ and η to finally obtain an energy contour plot as shown in figure 20.7. In this figure, the innermost contour correspond to the interaction energy of -19 MeV and the outermost to -1 MeV. This is a unique plot depicting the interaction energy characteristics of a positron entrapped within the proton core.

Figure 20.7

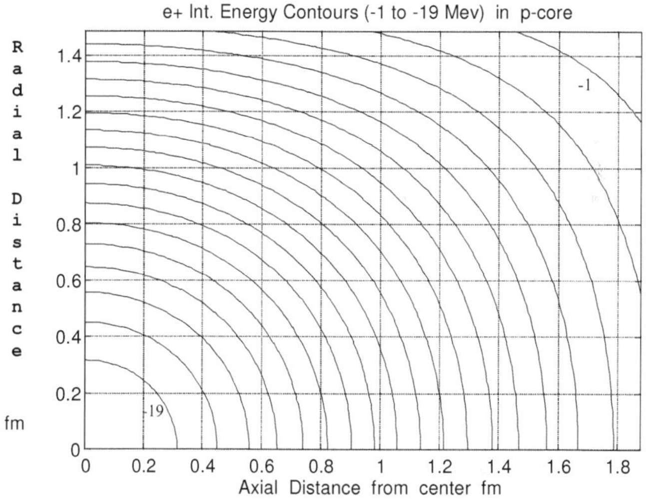

From this interaction energy data we can even compute the magnitude and direction of force experienced by the positron when its center is located at any point within the proton core. For example, the positron experiences maximum radial inward force of about 2600 Newton at a radius of 1.2 fm and a maximum radial outward force of about 600 N at a radius of 3 fm. A plot of positron inward radial force within the proton core is shown at figure 20.8. Similarly, in axial direction, the positron experiences a maximum inward force of about 1400 N at a distance of 1 fm from the center of the nucleon core. From the above mentioned data of the interaction energy and the associated force field acting on the positron entrapped within the nucleon core, we can make an estimate of the orbit on which the positron will move. With an estimated 2 MeV energy emitted out, the remaining amount of interaction energy is converted into kinetic energy of the entrapped positron. From dynamic considerations, orbital velocity of the positron is estimated at about 0.999 c, and the corresponding estimated orbital frequency is $N_r = 5.4533 \times 10^{22}$ Hz.

Figure 20.8

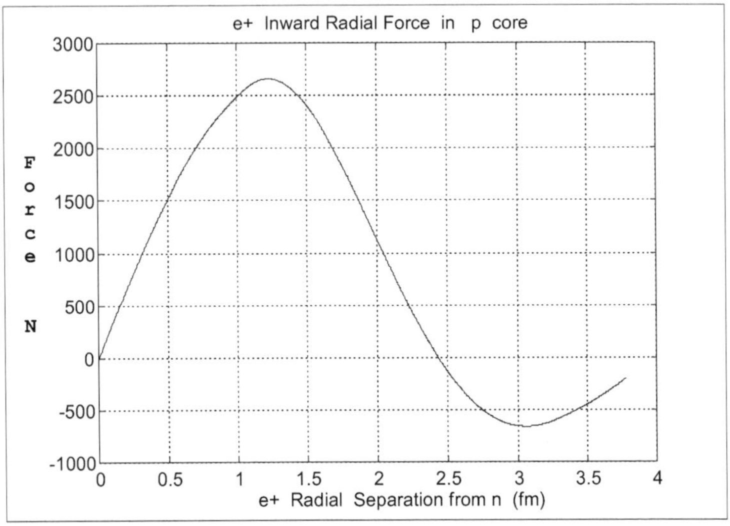

The major axis of this elliptical orbit is estimated at 2a=1.8384 fm and the minor axis 2b =1.6556 fm. The axis of elliptical orbit passes through the center of nucleon core and is inclined to the core axis at an angle of about 47.622 degree. Further, since the force field experienced by the positron is not exactly central, the positron orbit is also expected to precess around the nucleon core axis. The positron moving around the center of nucleon core on the above mentioned orbit will obviously produce a magnetic moment, the effective component of which along the core axis is,

$$\mu_p = eN_r . \pi ab . \mathrm{Cos}(\theta) = 1.407 \times 10^{-26} \ \mathrm{Am}^2 \qquad (20.32)$$

This is the familiar anomalous magnetic moment of the proton. Therefore, we may conclude that the proton consists of a positron entrapped inside a nucleon core through strong interaction and moving around the nucleon core center in elliptical orbits with fixed inclination to the core axis and bounded within a sphere of about one fm radius.

20.5 The Neutron

The neutron is known to be an uncharged nucleon with a negative magnetic moment. As such, it is quite reasonable to presume that the neutron may consist of a proton with an electron superposed over it in the outer periphery through strong interaction. We may therefore, examine the strong interaction characteristics of the electron core with the nucleon core. Since all strain components of the electron core are of opposite sign to those of the positron core, we can derive the interaction characteristics of the electron core from those of the positron core by simply changing the sign of the interaction energy. In this regard, let us consider the radial interaction of an electron core with a nucleon core for which the interaction energy will be given by the curve of figure 20.6 when vertically inverted about the zero base line. This shows that the electron will yield negative interaction energy when its center is located in the outer periphery of the nucleon core, within a cylindrical shell of radius 1.7 fm to 2.7 fm. Similarly, we can deduce from figure 20.8 that the electron located within a cylindrical shell of radius 2.3 fm to 2.7 fm of the nucleon core will experience an inward force of up to 650 N.

From the above mentioned data of the interaction energy and the associated force field acting on the electron entrapped in the outer periphery of the nucleon core, we can make a fairly good estimate of the elliptical orbit on which the electron will move. From dynamic considerations the orbital velocity of the electron is estimated at 0.137 c, with corresponding orbital frequency $N_r = 8.2\times10^{21}$ Hz. The major axis of this elliptical orbit of the electron is estimated at 2a=5.2 fm and the minor axis 2b=4.8 fm. The axis of the elliptical orbit passes through the center of nucleon core and is inclined to the core axis at an estimated angle of θ=22.5 degree. Further, since the force field experienced by the electron is not exactly central, the electron orbit is also expected to precess around the nucleon core axis.

The electron moving around the center of nucleon core on the above mentioned orbit will obviously produce a magnetic moment, the effective component of which along the core axis is,

$$\mu_{ne} = - e\, N_r\, \pi\, a\, b\, \mathrm{Cos}(\theta) = - 2.377 \times 10^{-26} \ \mathrm{Am}^2$$

However, this is just the component of magnetic moment that is attributed to the electron orbiting in the outer periphery of the nucleon core. We need to combine this component of magnetic moment with that of the positron μ_p orbiting deep inside the nucleon core. Therefore the total magnetic moment of the neutron is obtained as,

$$\mu_n = \mu_p + \mu_{ne} = 1.407\times 10^{-26} - 2.377\times 10^{-26} = - 0.970 \times 10^{-26}\ \mathrm{Am}^2 \quad (20.33)$$

This is the familiar anomalous magnetic moment of the neutron. Therefore, we may conclude that the neutron consists of an electron entrapped inside the outer periphery of a proton through strong interaction and moving around the nucleon core center in elliptical orbits with fixed inclination to the core axis.

Figure 20.9

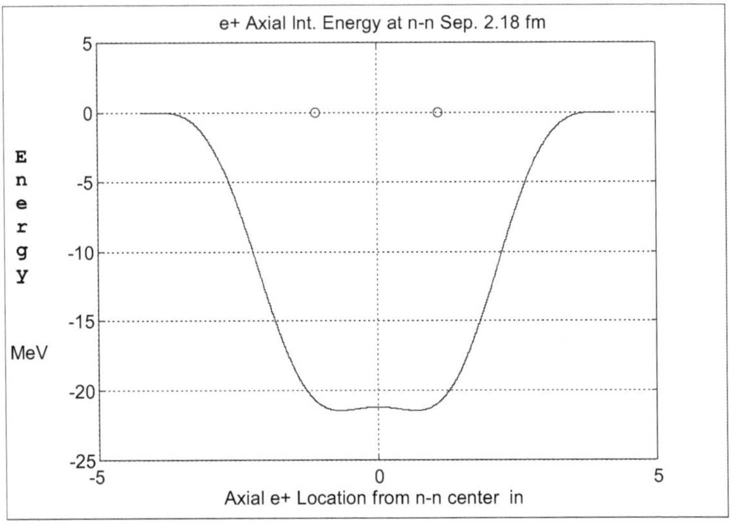

20.6 The Proton Neutron Bonding

As indicated above, the presence of an electron and a positron plays a major role in the bonding of nucleons together since there must be a valid mechanism available for emitting out a portion of the released interaction energy out of the system. The interacting nucleon cores by themselves do not appear to possess any such mechanism. As seen from figures 20.5 to 20.8, the electron and positron entrapped within specific regions of the nucleon core, display characteristic negative interaction energies. However, during strong interaction of two nucleon cores, the negative interaction energies of the positron or electron get enhanced in the common or superposed region of the interacting nucleon cores as shown in figures 20.9 and 20.10.

Hence, the entrapped electron or positron will tend to get shifted towards such enhanced negative interaction energy regions and acquire high kinetic energies in the process. When the high kinetic energies thus acquired are found to be not commensurate with their sustained trajectories, a portion of such kinetic energy is emitted out as mesons or neutrinos. Thus, it is the presence of electrons and positrons entrapped within the interacting nucleons that facilitates the emission of a portion of the interaction energy and hence ensures the bonding of the interacting nucleons.

Figure 20.10

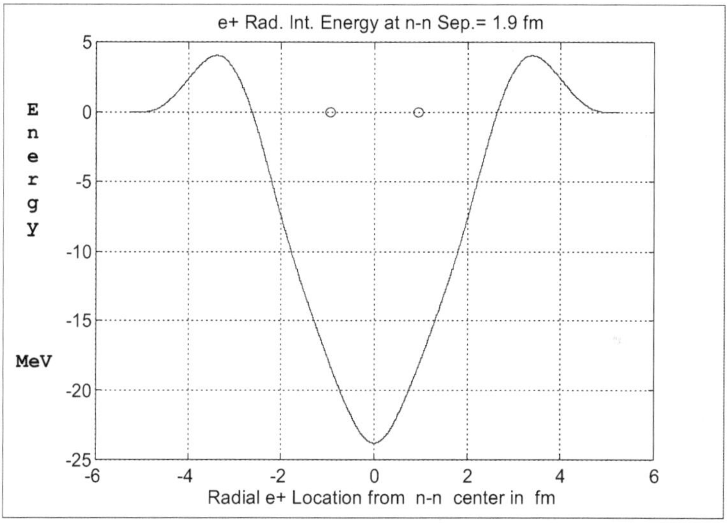

In an axial proton-neutron bond forming 'deuteron', the orbiting electron in the neutron core will tend to get axially shifted towards the common overlap region of the interacting cores and will gain some kinetic energy in the process. Since the axially interacting nucleon cores tend to vibrate at a frequency of about 1.8×10^{21} Hz, the orbiting electron will emit out a part of its kinetic energy to synchronize its motion with the nucleon oscillations. During each cycle of their oscillations, when the nucleon cores are closest together at a separation distance of about 2.18 fm, the electron will tend to be in their mid section. The electron orbit may even keep shifting from one nucleon core to the other after each cycle of rotation.

As shown in figure 20.4b, when two nucleon cores are radially separated by a distance of 4.3 fm or more, they will tend to repel each other. Therefore, in radial proton-neutron bonding, the presence of an orbiting electron in the outer periphery of the neutron will neutralize this nucleon core repulsion and assist in bringing the neutron and proton closer. As pointed out earlier, two nucleon cores when coupled radially, will tend to vibrate at a frequency of about 3.6×10^{22} Hz with a minimum separation of about 1.9 fm between their axes. The combined radial interaction energy plot of a positron interacting with two nucleon cores radially separated by 1.9 fm is shown at figure 20.10. Here again, each orbiting positron from the interacting neutron and proton, will tend to synchronize its motion with radial vibrations of nucleon cores by radiating out a portion of its kinetic energy. The amount of energy thus radiated out by the two positrons will become the effective bond energy of this p-n radial coupling. This bond energy is relatively a very small fraction of the total kinetic energy available in the system that sustains vibrations and rotations of constituent nucleons and orbital motion of the entrapped positrons.

There is one very important feature of the synchronous orbital motion of positron within a radial p-n coupling. The combined orbit of positron around two radially coupled nucleon cores will be of the shape of figure of '8'. If the sense of positron motion around one nucleon core is clockwise, it will be anti-clockwise around the second core. This fact might be responsible for creating the general impression that conventional spins of two nucleons are parallel for their axial coupling and anti parallel for their radial coupling. Whereas the positrons will still be orbiting within the inner regions of each nucleon core, the lone electron from the neutron will however, keep orbiting around the outer periphery of each core in the figure of '8'. Of course, the electron will cross the overlapping region of the two cores when they are farthest apart in the oscillation cycle while the positrons will cross the overlapping region when the cores are closest. One important point to be noted here is that while the presence of electron and positron is crucial for facilitating the axial or radial bonding between the proton and the neutron by emitting out a portion of the interaction energy, their presence is no longer essential for maintaining an established bond. The subsequent orbital motions of the electron and the positron are the dynamic consequences of the interaction forces.

Finally, it may be concluded that the detailed understanding of the strong interactions among nucleons, positrons and electrons can be of great value in the planned development of nuclear fusion processes and controlled fusion devices. Specifically, it is important to note that all interacting nucleons must be axially aligned with the use of appropriate magnetic fields, to facilitate their axial and radial interactions. For the production of Deuterons, it may be necessary to axially align the bulk (or beams) of neutrons and subject them to high frequency axial vibrations. Simultaneously however, a beam of high energy positrons will need to be injected along the axial direction of the axially aligned neutrons. On the other hand, to facilitate the radial interaction of axially aligned pairs of Deuterons, a beam of high energy electrons will need to be injected along the radial direction while the Deuterons are subject to high frequency radial vibrations. Extensive further studies are called for in this field.

21

Stable and Unstable Elementary Particles

21.1 Strained State in the Elastic Space Continuum

When any region of the Elastic Space Continuum is subjected to some sort of deformation, a strain field may be said to have developed in that region. This strain field can be fully defined, if the displacement vector **U** is completely determined as a function of space and time coordinates over the whole region of the Continuum under deformation. The displacement vector components u^i can be completely determined from the detailed solution of the equilibrium equations,

$$g^{jj}u^i_{,jj} = (1/c^2)\, \partial^2 u^i/\partial t^2 \tag{21.1A}$$

$$\nabla^2 \mathbf{U} = (1/c^2)\, \partial^2 \mathbf{U}/\partial t^2 \tag{21.1B}$$

Hence, the detailed study of any deformed or stressed region of the Space Continuum primarily involves the detailed solution of the equilibrium equations subject to appropriate boundary conditions. For different sets of boundary conditions, the given partial differential equations will yield different unique solutions. In some special cases, the method of separation of variables is applicable for the solution of equilibrium equations in a given coordinate system.

General Boundary Conditions. Let V be the total volume and Σ the outer boundary surface of a particular region of the Elastic Space Continuum under deformation. The general boundary conditions that must be satisfied by the displacement vector components u^i obtained from the solution of equilibrium equations (21.1), may be listed as:

- The displacement components u^i must vanish at the boundary Σ and must remain finite and continuous within this boundary.
- The strain components and the energy density must be finite and continuous within the boundary Σ of the region under consideration.
- The total strain energy within the entire volume V must be finite and remain constant with time in the absence of any external interaction.

21.2 Strain Bubbles or Elementary Particles

A closed region of the Elastic Space Continuum with boundary surface Σ, wherein the displacement vector components u^i obtained from the solution of equilibrium equations (21.1), satisfy the above mentioned boundary conditions and contains a finite amount of energy stored in its strain field, may be called a 'Strain Bubble' or an elementary particle. From the nature of general boundary conditions and the equilibrium equations, it turns out that all valid solutions for displacement vector components u^i are functions of space-time coordinates representing various types of strain wave oscillations. That is, all 'Strain Bubbles' contain a constant finite amount of total strain energy and essentially consist of various strain wave oscillations within a specific boundary surface Σ of the Elastic Space Continuum. The mode of strain wave oscillations constitutes the main distinguishing features of various types of strain bubbles.

Strain Bubble Formation. If a certain finite amount of energy is somehow transferred to a particular region of the Elastic Space Continuum such that it attains a stable configuration, then a sustained strain field will develop in that region. One of the crucial conditions for the formation and stability of such strain bubbles is the time invariance or conservation of the total strain energy content of the strain field within the specified bounded region of space. Although the strain components will always be functions of space & time coordinates, yet the strain energy density may or may not be time invariant. Even with such constraints, a large number of different varieties of strain bubbles can exist or coexist within the Elastic Space Continuum.

21.3 Stability Criteria for Elementary Particles

Most of the elementary particles consist of either single strain bubbles or a composite of two or more strain bubbles bound together through strong interaction. Therefore, the first criteria for the stability of elementary particles that are composites of two or more strain bubbles is that the binding energy of the strain bubbles must be high in comparison to the kinetic energies of individual constituent strain bubbles. A further condition for the stability of such composite particles is that each of the constituent strain bubbles must itself be stable. However, any such composite particle can break up if sufficient energy is pumped into it from external sources through collisions or strong interactions.

All strain bubbles, that can exist in a stable configuration for only a short but measurable duration of time, can be identified with unstable elementary or composite particles. At the end of their life time, such unstable particles invariably get transformed into some other lower energy particles or wavelets and may release a portion of their energy content as kinetic energy of other interacting particles. All other strain bubbles that can exist in a stable configuration for a long long time can be regarded as stable elementary or composite particles. The essential notion of stability of strain bubbles concerns their existence in a stable configuration in the absence of any strong interaction with other strain bubbles. Therefore, we need to highlight the circumstances, or more precisely the structural peculiarities that distinguish the stable elementary particles from the unstable ones.

As noted above, the strain field within a strain bubble is defined by the displacement vector components u^i obtained from the solution of equilibrium equations (21.1) subject to the boundary conditions. The strain components are always functions of space & time coordinates. If the strain energy density W, obtained from the sum of squares of all strain components, is completely time invariant function of space coordinates, the associated strain bubble can be regarded as a stable elementary particle. However, if the strain energy density W oscillates periodically, within finite limits, such that the total energy content within the boundary of a strain bubble remains constant or time invariant, that strain bubble can still be regarded as a stable elementary particle.

There are two additional constraints on the above mentioned stability criteria of elementary particles. For a particular configuration of a strain bubble, apart from the energy density being time invariant, the total energy content must be a minima from among the energy contents of various other alternative strain bubble configurations. Further, for a stable configuration of a strain bubble, the spatially bounding functions occurring in the solutions of displacement vector components u^i, must be limited to their first zeros or nodes. That is because the strain bubble configurations that extend to higher nodes of their bounding functions will constitute higher modes of oscillations within the strain bubble and will be comparatively unstable.

As an example of a most stable strain bubble, let us consider the configuration of a bare nucleon core which is common for both the proton as well as the neutron. The displacement vector components u^i, in cylindrical coordinates ρ, ϕ, z are (with $u^z = 0$) given as,

$$u^\rho = \pm A_1.e\kappa.\ J_1(x).\ \mathrm{Cos}(qz).\ \mathrm{Cos}(\kappa ct) \qquad (21.2A)$$

$$u^\phi = \pm A_1.e\kappa.\ J_1(x).\ \mathrm{Cos}(qz).\ \mathrm{Sin}(\kappa ct) \qquad (21.2B)$$

where, A_1 is a dimensionless number, $x = (\kappa^2 - q^2)^{1/2}\rho$, $J_1(x)$ is the Bessel function and the boundary surface is given by $-\pi/2 \le qz \le \pi/2$ and $0 \le x \le \alpha_1$ with $J_1(\alpha_1) = 0$ or $\alpha_1 = 3.832$. Here, κ is the wave number of strain wave oscillations. The Strain energy density for this particle works out to be,

$W_1 =$

$$\frac{A_1^2 e^2 \kappa^2}{2\varepsilon_0}\left[(\kappa^2 - q^2)\left\{(J'_{\mathrm{size7}\,1}(x))^2 + \frac{J_1^2(x)}{x^2}\right\}\mathrm{Cos}^2(qz) + J_1^2(x)\{\kappa^2\,\mathrm{Cos}^2(qz) + q^2\,\mathrm{Sin}^2(qz)\}\right]$$

Since this energy density is completely independent of time, the strain bubble represented by equations (21.2) is expected to be most stable. However, if we increase the spatial extension of this strain bubble beyond $x = \alpha_1$ to say $x = \alpha_n$ where $\alpha_n > \alpha_1$ and $J_1(\alpha_n) = 0$, then this strain bubble with higher total energy content, will no longer be as stable. Such strain bubbles, with higher mode strain wave oscillations and with or without superposed charge particles, constitute a very important class of unstable elementary particles called hyperons.

21.4 Unstable Elementary Particles

As noted above, the first category of unstable elementary particles consists of higher mode spatial extensions of all stable particles. For example the higher mode unstable particles in the proton series will have their nucleon cores extended to higher nodes than first. That is, their radial boundary surface may extend to $0 \le x \le \alpha_2$ with $J_1(\alpha_2) = 0$ and $\alpha_2 > \alpha_1$ or $0 \le x \le \alpha_3$ with $J_1(\alpha_3) = 0$ and $\alpha_3 > \alpha_2 > \alpha_1$. Also, their axial boundary surface may extend to $-3\pi/2 \le qz \le 3\pi/2$. This is also true for all other stable particles like positron and electron. The natural decay of this category of unstable particles will proceed from the higher mode to the lower mode up to the lowest stable mode of that strain bubble.

The second category of unstable elementary particles consists of those strain bubbles, for which the components u^i are valid solutions of equilibrium equations (21.1), but the corresponding strain energy density is not time invariant. Three important solutions in this category are:

$$u^\rho = \pm A_2.e\kappa.\ J_1(x).\ Cos(qz).\ Cos(\kappa ct) \quad \text{with } u^\phi = 0 \ \&\ u^z = 0 \qquad (21.3)$$

$$u^\phi = \pm A_3.e\kappa.\ J_1(x).\ Cos(qz).\ Sin(\kappa ct) \quad \text{with } u^\rho = 0 \ \&\ u^z = 0 \qquad (21.4)$$

and $u^z = \pm A_4.e\kappa.\ J_0(x).\ Cos(qz).\ Sin(\kappa ct)$ with $u^\rho = 0$ & $u^z = 0$ (21.5)

where, A_2, A_3, A_4 are dimensionless numbers, $x=(\kappa^2-q^2)^{1/2}\rho$ and the boundary surface is given by $-\pi/2 \le qz \le \pi/2$ & $0 \le x \le \alpha_1$ with $J_1(\alpha_1) = 0$.

The strain energy density in these strain bubbles oscillates with time, thus rendering them unstable, even though the total strain energy remains time invariant. These strain bubbles are capable of strong interactions with other strain bubbles containing similar displacement vector components u^i. From a detailed study of their interactions, these strain bubbles are likely to be identified with the 'cores' of different mesons. One important category of strain bubbles with highly oscillating strain energy density could be termed 'strain energy resonances' which are identified with highly unstable elementary particles.

22

Fundamental Nature of Gravitation

22.1 Current Theories of Gravitation

Gravitation is a natural phenomenon by which all objects with mass attract each other, and is one of the fundamental forces of physics. Newton's theory of universal gravitation is a physical law describing the gravitational attraction between bodies with mass. In 1687, Sir Isaac Newton published Principia, which hypothesizes the inverse-square law of universal gravitation. As per this law, every point mass attracts every other point mass by a force pointing along the line intersecting both points. This force is proportional to the product of the two masses and inversely proportional to the square of the distance between the point masses, as

$$F = G.(m_1 * m_2)/r^2 \qquad (22.1)$$

where G is the universal gravitational constant, approximately equal to 6.67×10^{-11} N m^2/ kg^2.

For interacting bodies with spatial extent, the gravitational force between them is calculated by summing the contributions of the notional point masses which constitute the bodies. It can be shown that an object with a spherically-symmetric distribution of mass, exerts the same gravitational attraction on external bodies as if all the object's mass were concentrated at a point at its center. Gravitational potential field V(r) of a spherically symmetric body of mass M is defined at a distance r from its center as,

$$V(r) = -G*M/r \qquad (22.2)$$

The gravitational potential V is an interaction parameter, the magnitude of which depicts the amount of interaction energy released from the inherent field energies of the interacting masses when a unit mass is brought from infinite separation to distance r from mass M.

Newton's law of gravitation is similar to Coulomb's law of electrical forces, which gives the magnitude of electrical force between two charged particles. Both are inverse-square laws, in which force is inversely proportional to the square of the distance between the particles. Coulomb's Law has the product of two charges in place of the product of the masses, and the electrostatic constant in place of the gravitational constant. This similarity points to a possibility that the gravitational and electrostatic interactions might have something in common at the most fundamental level.

However, the major theoretical concern of Newtonian Gravitation is that the mediator of gravitation could not be identified so far. That is, there is no fundamental physical mechanism that could explain the action of one massive body over the other body separated by a large distance through the void or empty space. But a similar theoretical concern also existed for the action of one charged particle over the other charge separated by a large distance through the void or empty space. Newton's own remarks in this regard are relevant,

"*That one body may act upon another at a distance through a vacuum without the mediation of anything else, by and through which their action and force may be conveyed from one another, is to me so great an absurdity that, I believe, no man who has in philosophic matters a competent faculty of thinking could ever fall into it.*"

Gravitation in General Theory of Relativity. General theory of relativity (GR) is the geometric theory of gravitation propounded by Albert Einstein. In GR, the effects of gravitation have been ascribed to spacetime curvature instead of a force. The starting point for general relativity is the equivalence principle, which equates free fall with inertial motion. In GR, it is assumed that spacetime is curved by matter, and that free-falling objects are moving along locally straight paths, called geodesics in curved spacetime. Through Einstein's field equations (EFE), the presence of matter and energy content in a certain region of space is related to the curvature of spacetime continuum.

The EFE are a set of 10 simultaneous, non-linear, partial differential equations and their solutions yield the components of metric tensor of spacetime. A metric tensor describes the geometry of spacetime. A non-zero value of the Riemann tensor, computed from the metric tensor components, is said to describe the curvature of spacetime. The geodesic paths for a spacetime are also calculated from the metric tensor. An important solution of EFE known as the Schwarzschild solution, model the static gravitational field of a spherically symmetric body of mass M.

However, as already discussed in chapters 8 and 9, the GR is founded on ill-conceived notions of spacetime continuum and the so-called curvature of spacetime. Imagine a time coordinate axis extending from minus infinity to plus infinity with the present time marked at point T_p. The spacetime is assumed to be a 4-D block with 3-D physical space existing at each and every point of the time axis and not just at the present time point T_p. The matter and energy content in a certain region of physical space is assumed to influence the metric of whole spacetime, that is including the past, present and future regions of spacetime as per EFE. Further, there is no physical mechanism which could justify the postulate of matter influencing the metric of spacetime. Actually, spacetime is just a mathematical notion which has no physical existence.

In reality, of course, GR is just a mathematical model in which the Riemannian 4D space-time manifold is being used as a differential scale template for getting the trajectories of objects as geodesic curves. For obtaining trajectories of objects moving in 3D physical space, we can first obtain a differential scale 4D manifold XYZ-T as a template such that the Newtonian trajectories in the given gravitational field appear as geodesic curves in this template manifold. To do this, the differential scale or the metric of this template manifold will have to be correlated with the mass M of the gravitating body.

Now, to obtain the trajectory of any other object in the given gravitational field, we can mark the initial starting position of the object in the template manifold and then compute the trajectory as a geodesic through that position. However, we will have to adjust the differential scale or the metric coefficients of this template manifold according to the mass M of the gravitating body. This is precisely what is being done through EFE in the GR model of gravitation. Additionally, certain proportionality and integration constants are arbitrarily fixed or adjusted in such a way that the gravitational acceleration in the vicinity of earth is matched with the corresponding value obtained from the Newtonian theory of gravitation.

Thus, it is clear that GR is just a mathematical model used for obtaining the trajectories of objects as geodesic curves. However, the supporters of GR did attempt to elevate this mathematical model to the status of a physical theory by 'assuming' the 4D space-time manifold to be a physical spacetime continuum and also 'assuming' that the mass-energy content of a gravitating body 'somehow' controls the metric of this physical entity. Once we realize that 4D spacetime manifold is just an abstract mathematical notion and not a physical entity, then it is quite a simple matter to understand that GR is just a mathematical model and not a physical theory.

22.2 Coulomb Interaction and Electrostatic Field Energy

Once we realize that the geometric model approach of GR is fundamentally wrong for understanding the gravitational phenomenon, we must explore the possibility of removing the 'theoretical concerns' of Newtonian Gravitation. For this, we need to understand the fundamental physical mechanism of gravitational interaction. Since we have already noted the similarity of the gravitational interaction between two massive bodies and the electrostatic interaction between two charged bodies, we need to further explore the electrostatic phenomenon for obtaining some valuable insight into the gravitational phenomenon.

As discussed earlier, the electrostatic field of an electron type charge particle consists of phase waves with radially decaying amplitude, propagating inwards from infinity to the core boundary, at the speed of light 'c'. The electrostatic field of positron type charge particle consists of phase waves with radially decaying amplitude, propagating outwards from the core boundary to infinity, at the speed of light 'c'. We may combine the direction of propagation of the phase waves with the rms value of their amplitudes to get an effective field strain tensor **S** defined at each and every point of the electrostatic field. During the process of interaction of charges, their electrostatic wave fields will get superposed through the superposition of the effective field strain tensors **S**. Strain components in the superposed field will then be given by the effective combined strain tensor $\mathbf{S_{sup}}$.

To compute Coulomb interaction between two opposite charges [1,2], we have to superpose the effective strain tensors $\mathbf{S_1}$ and $\mathbf{S_2}$ of their respective wave fields and then compute the combined strain components $\mathbf{S_{sup}}$. The total energy E_{sup} of their combined fields can then be computed from the superposed strain components. The difference $[E_{sup} - (E_1 + E_2)]$ is termed as the interaction energy, E_{int}. For two dissimilar charges the combined field energy E_{sup} will reduce and the interaction energy will be termed -ve. The negative interaction energy implies that due to the superposition of fields, part of the initial total field energy of the system of interacting charges is released by the system and may get transformed to their kinetic energy or to some other form of energy.

The negative (radially inward) electrostatic phase wave field of the electron accommodates about 35% of the total mass energy of the electron. That is, about 175 KeV energy is contained in the electrostatic wave field surrounding the core region of the electron. However, the energy density in this electrostatic wave field decreases in proportion to $1/R^4$ from the boundary of the core outwards (R>1.61 fm). When we consider the electrostatic interaction of a proton (or a positron) and an electron, the total electrostatic field energy contained in their respective wave fields will be about 350 KeV ($E_1 + E_2$). In the process of their interaction, their respective wave fields will get superposed, with the result that:

(a) The total energy E_{sup} of their combined fields can then be computed from the superposed effective strain tensor field $\mathbf{S_{sup}}$. As long as the separation distance between the proton and the electron remains greater than 10 fm (10^{-15} m), the total energy E_{sup} in their combined fields will be greater than 200 KeV. This is a very significant result.

(b) The superposed electrostatic strain wave field will no longer be propagating in any one direction (radially inwards or radially outwards) but will consist of their superposed mixture which may be termed as a *ripple wave field.* This ripple wave field will no longer behave as an electrostatic phase wave field of a charged particle even though a significant amount of total energy E_{sup} is still contained in this field.

22.3 Ripple Wave Field Energy of Atoms and Molecules

Most of the neutral atoms and molecules consist of a large number of proton-electron pairs held together in specific bound states. The ripple wave fields of different proton-electron pairs in an atom or molecule will further get superposed to yield a combined superposed ripple wave field of the neutral particle. We may refer to this combined superposed ripple wave field of neutral atoms and molecules as a neutral ripple wave field with combined strain tensor field $\mathbf{S_f}$ and corresponding total strain energy content E_f . If certain neutral particle consists of N pairs of protons and electrons bound together in a specific configuration, then the total strain energy contained in the neutral ripple wave field of this particle will be given by,

$$E_f = N*350 - [\text{total binding \& kinetic energy of N electrons}] \text{ KeV} \qquad (22.3)$$

However, since the minimum separation between any proton and electron in a neutral particle is much more than 10 fm, the minimum content of total strain energy E_f contained in the neutral ripple wave field will be much more than N*200 KeV. Further, since the mass of N pairs of protons and electrons is proportional to N, the total strain energy E_f contained in the neutral ripple wave field of any neutral particle will be proportional to the mass of that particle.

22.4 Gravitation as a Ripple Wave Field Interaction

Let us now consider whether two (or more) neutral particles, with such high energy content E_f in their neutral ripple wave fields $\mathbf{S_f}$, can interact in any way. That is, we have to examine the superposition of two or more neutral ripple wave fields $\mathbf{S_f}$ to determine their interaction. During any such interaction, the effective strain tensor fields $\mathbf{S_f}$ of all neutral particles in the region will get superposed through 'tensorial' addition of $\mathbf{S_{f1}}$, $\mathbf{S_{f2}}$ etc. to get a resultant strain tensor $\mathbf{S_g}$ at all field points as,

$$\mathbf{S_g} = \mathbf{S_{f1}} + \mathbf{S_{f2}} + \mathbf{S_{f3}} + \ldots + \mathbf{S_{fn}} \qquad (22.4)$$

The strain components at each and every field point are now given by the resultant strain tensor field $\mathbf{S_g}$. From these field strain components, we can compute the strain energy density and the total energy E_g contained in the whole region of the superposed neutral ripple wave fields. Finally, the interaction energy E_{int} of all neutral particles in the region will be given by

$$E_{int} = E_g - [E_{f1} + E_{f2} + E_{f3} + \ldots + E_{fn}] \qquad (22.5)$$

This interaction energy is expected to be negative, implying that such an interaction among neutral particles will provide an attractive force between such particles and bind them together. Such an interaction among neutral matter particles is identified with the *gravitational* interaction.

22.5 Physical Mechanism of Newtonian Gravitation

As per Newtonian theory of gravitation, every mass particle attracts every other mass particle by a force proportional to the product of the two masses and inversely proportional to the square of the distance between them. Until now there was no fundamental physical mechanism available to explain the action of one massive body over the other body separated by a large distance through the void or empty space. The interaction between two or more neutral matter particles through superposition of their neutral ripple wave fields, as discussed above, provides an excellent fundamental physical mechanism for explaining the gravitational interaction.

Thus, the fundamental basis of gravitational interaction can be traced to the electrostatic phase wave field of charged particles. This field is characterized by unidirectional (radially inwards or outwards) propagation of phase waves with radially decaying amplitude and may be represented by an *effective* strain tensor field **S**. The strain energy density and the total energy content of the electrostatic field is completely determined by the effective strain tensor field **S**. However, the fundamental unit cell for gravitational interaction is the neutral pair of two opposite charges and the basic particle for gravitational interaction is the neutral atom or molecule which may consist of a large number of unit cells held together in a bound state.

The phase wave field of such neutral atoms or molecules is characterized by neutral ripple wave field obtained by superposition of all phase wave field contributions from all constituent charges and may be represented by the resultant strain tensor field $\mathbf{S_f}$. The gravitational interaction of all neutral atoms and molecules is characterized by the superposition interaction of their ripple wave fields, represented by strain tensor fields $\mathbf{S_f}$, to yield a resultant gravitational ripple wave field given by strain tensor $\mathbf{S_g}$ (eqn. 22.4). The strain energy density computed from the strain tensor field $\mathbf{S_g}$, leads to a net (negative) interaction energy E_{int} (eqn. 22.5) that gets released from the system of gravitationally interacting particles.

It may therefore be emphasized that all neutral atoms, molecules and composite particles are surrounded by a characteristic ripple wave field, with radially decaying amplitude, that contains a significant amount of strain energy. The gravitational interaction between such particles is mediated through the superposition of their ripple wave fields, resulting in a fine re-adjustment of field strain energy distribution and release of a small fraction of the field energy as interaction energy E_{int}. Due to a high strain energy density in the surrounding ripple wave field of all material particles, it is quite possible that the motion of elementary particles such as photons may get influenced by the gravitational field of all matter bodies even if these elementary particles do not contribute to the gravitational fields.

Validity of Equivalence Principle. As noted earlier, a neutron contains one positron embedded in the central region and one electron embedded (orbiting) in the peripheral region of the nucleon core through strong interactions. Since the separation distance between the positron-electron pair is less than 3 fm, the residual strain energy contained in their ripple wave field is expected to be much less than the corresponding strain energy contained in the atomic charge pairs. Hence the contribution of neutrons to the overall ripple wave field of neutral matter particles is likely to be much less as compared to the contribution of atomic proton-electron pairs.

In general terms, the Equivalence Principle implies the equivalence of gravitational mass and inertial mass. But we have just seen that in the buildup of ripple wave gravitational field, the contribution of neutrons per unit mass will be much less than the contribution of atomic proton electron pairs (per unit mass). Hence, the equivalence of gravitational mass and inertial mass can be ensured only if the ratio of neutrons to protons is strictly maintained constant for all neutral matter particles. This, however, is only broadly true and not strictly true. Hence, it appears that the Equivalence Principle is neither a fundamental principle of nature nor is it strictly valid for all forms of matter. This observation may have a significant impact on our overall world view and cosmology. Although with the collapse of relativity theories, the fantasy of the Big Bang and Black Holes turns out to be absurd, it is quite possible that even the so-called 'Neutron Stars' may turn out to be just high density (metallic) plasma stars. Further, since the gravitational field of a neutron particle is expected to be significantly weaker than the gravitational field of an atomic proton-electron pair, the gravitational collapse of a big star may in fact get reversed with the formation of a so-called 'Neutron Star'.

23

Summary: Call for a Paradigm Shift

23.1 Current Paradigm

A set of beliefs, dogmas, postulates and axioms, along with concepts and notions derived from particular interpretation of different physical observations, constitute the paradigm of fundamental physics. Under the current paradigm, it is generally believed that a fundamental theory of Physics need not be based on logical foundation and need not contain causal linkages for physical explanation. However, it is a fact that the current paradigm does not provide logical foundations to the mathematical models which are being projected as physical theories. A paradigm shift – a significant change in the paradigm is needed . Salient features of the current paradigm that need to be revised, are as follows:

(a) ***Empty Space.*** The void between material particles represents the notion of empty space with the geometrical attribute of volume. There is no distinction between the abstract notion of coordinate space with a specified metric and physical space. No physical properties are believed to be associated with empty space, even though it can get curved.

(b) ***Spacetime.*** All physical phenomena are believed to occur in a four dimensional world of spacetime. The dimension of time is treated at par with three spatial dimensions in the spacetime. All dynamic phenomena occurring in space can be represented through the geometrical constructs in a four dimensional spacetime manifold. Physical reality of spacetime continuum implies a belief in four dimensional block view of universe, in which three dimensional physical world is assumed to exist at all points on the time axis. Additionally, the spacetime continuum is believed to get curved under gravitation. The curved space representing gravitational field is also believed to store energy. The notion of *curvature of space* and spacetime is a keystone in the current paradigm of Relativity.

(c) Relativity

i. Space and time measurements are *considered* relative and not absolute. A particular value of distance and time measurement will *appear* to be different for different observers in relative uniform motion.

ii. The speed of light is *believed* to be a constant 'c' for all observers in relative uniform motion. The relativity of space and time stems from this postulate of constant 'c' for all observers in relative uniform motion.

iii. All reference frames in relative uniform motion are considered *equivalent* in all respects and none of them can be regarded as preferred or unique or absolute in any sense that could be practically distinguished. Establishment of any fixed, absolute or universal reference frame is *considered* impossible.

iv. All forms of matter and energy are *believed* to influence the metric of spacetime without any physical mechanism.

v. The metric of space is *believed* to be a physical property of empty or free space.

vi. The 'fabric' of spacetime is *believed* to be so deformable that the metric of curved spacetime can also lead to mathematical singularities which are interpreted as physical Black Holes.

(d) ***Big Bang Cosmology.*** The Big Bang Model is a broadly accepted theory for the origin and evolution of our universe. It postulates that the present universe has expanded from an initial spacetime singularity into the vast cosmos we currently inhabit.

(e) ***Expansion of space in the Expanding Universe.*** The metric expansion of space is the averaged increase of measured distance between objects in the universe with time. It is an intrinsic expansion – i.e. it is defined by the relative separation of parts of the universe and not by motion "outward" into preexisting space. Metric expansion is a key feature of Big Bang cosmology.

(f) ***Elementary particles.*** All elementary particles are believed to be point like and structureless particles that are not composed of smaller particles. The elementary particles are assumed to be endowed with special interaction characteristics. We do not know anything about the shape, size and internal structure of any of the elementary particles.

(g) ***Existence of Fields.*** A field in physics represents a physical quantity that could be associated to each point of space-time. A field is believed to contain energy and momentum. The physical existence of fields make the field concept a supporting paradigm of the entire edifice of modern physics. But how does a field physically ***exist*** at each point of the space-time? The physical mechanism that could support the existence of a field at each point of space and time, is neither understood nor being explored.

(h) ***Kinetic and Potential Energy.*** All particles of matter are assumed to be bestowed with specific amounts of kinetic and potential energies. Whereas the kinetic energy depends upon the motion, the potential energy depends upon the position of the particle. The sum of kinetic and potential energy of all particles within a closed system, remains constant invariant.

(i) ***Exchange theory of interactions.*** All particle interactions are believed to be mediated through the exchange of special force mediating particles called bosons. In particular, electromagnetic interaction among charged particles is believed to be mediated through the exchange of virtual photons.

(j) ***Probability waves.*** All microscopic particles of matter display a wavelike nature while in motion. The intensity of the wavelet is interpreted as the position probability density for the location of the micro particle. This interpretation implies that probability waves accompany all micro particles in motion.

(k) ***Uncertainty in the trajectory of micro particles.*** Consequent to the assumption of probability waves accompanying micro particles in motion, the description of their trajectory parameters becomes probabilistic or uncertain to some extent. This uncertainty in the description of trajectory parameters of micro particles has evolved into a fundamental principle of quantum mechanics.

(l) ***Weirdness of Nature.*** It is believed that the inherent weirdness of Nature does not permit us to mentally visualize the shape, size or structure of all micro particles and their associated fields.

23.2 The New Paradigm

(a) Mathematical representations supposed to describe physical reality are not sufficient to enable us to mentally visualize the associated physical phenomenon. Such mathematical representations must be supported by causal linkages, physical explanations and logical foundations, to adequately describe physical reality.

(b) It is not the inherent complexity of Nature that prohibits us from mental visualization of fundamental physical phenomenon, but our ignorance of Nature. Just like computer simulations, if we possess sufficient relevant information about certain physical phenomenon, we must be able to mentally visualize the same.

(c) To understand and comprehend physical phenomenon occurring in physical space or vacuum, we must first attempt to understand the fundamental nature of physical space or vacuum. The physical properties of permittivity, permeability and intrinsic impedance characterize the fundamental nature of empty or free space.

(d) Apart from a working mathematical model, a physical theory must incorporate causal linkages for logical explanation of associated physical phenomena.

(e) The popular notion of 'curvature of space' actually implies the 'deformation of space'. The deformation of space can be properly represented through a displacement vector and a strain tensor associated with each and every point of the space continuum.

(f) An absolute fixed reference frame, such as the International Celestial Reference Frame (ICRF) can be constructed in any closed volume of space with finite matter content. The concept of an absolute reference frame can be extended to the Universal Reference Frame which can be practically established just like the ICRF.

(g) In the mathematical model of General Relativity, an abstract notion of spacetime manifold has been used as a graphical template, with differential scaling of its space and time axes, to represent the particle trajectories as geodesic curves. The four dimensional spacetime continuum is not a physical entity in any sense but just an abstract mathematical notion.

(h) The uncertainty principle of Quantum Mechanics is strictly applicable for certain mathematical representations of the physical phenomenon and not to the phenomenon itself.

(i) A detailed analytical study of dynamic deformations in the space continuum, through the time dependent displacement vector field **U**, provides a more fundamental level of investigation into the workings of Nature, in comparison to the fields currently used for the purpose.

(j) All forms of energy and mass fundamentally exist as the strain energy of the dynamic deformation or strained state of the physical space continuum.

(k) All physical phenomena involving matter particles and fields, represented through various strained states of the space continuum and their mutual interactions, can be mentally visualized as well as mathematically modeled.

(l) Quantum Mechanics needs to be recast into Bohmian Mechanics after replacing the notion of probability waves with strain waves in physical space continuum. The kinetic energy of a particle in motion is the strain energy stored in the pilot waves accompanying the particle.

23.3 Matter Particles and Fields

(a) All electromagnetic phenomena, energy transportation processes and all wave motion that are believed to be occurring in empty space, are in fact occurring in the elastic space continuum with characteristic properties of permittivity ε_0 and permeability μ_0.

(b) The equilibrium equations of elasticity in the Elastic Space Continuum turn out to be identical to the Maxwell's vector wave equation for the electromagnetic field, as $\nabla^2\mathbf{U} = (1/c^2)\,\partial^2\mathbf{U}/\partial t^2$.

(c) A closed region of the Elastic Space Continuum in a permissible strained state, satisfying the equilibrium equations & boundary conditions, may be termed as a strain bubble.

(d) The total strain energy content E_0 of a strain bubble represents its rest mass m_0, through the well known mass-energy equivalence relation $E_0/c^2 = m_0$.

(e) The electromagnetic field in free space is just a dynamic strained state of the elastic space continuum. Whereas the temporal stress in the continuum corresponds to the electric field **E** in free space, the torsional stress in the space continuum corresponds to the magnetic field **B**.

(f) Interaction energy, or the conventional potential energy, of two interacting strain bubbles may be defined as the difference between the total strain energy with superposed strain fields and the sum of their separate strain field energies. A negative interaction energy or potential energy will imply the release of a portion of the total strain energy of the two interacting bubbles. The released energy may either transform into another strain bubble wholly or partly and emitted out of the system, or transform into kinetic energy of motion of the interacting strain bubbles.

(g) The clusters of pure and composite strain bubbles depicting 'elementary particles' are essentially characterized by their 'strain energy content', 'strain wave fields' if any and their interaction properties. In principle, there could be an infinitely large number of different types of strain bubbles occurring in the Elastic Space Continuum, that may be correlated with an equally large number of stable and unstable elementary particles.

(h) The electron and positron pair is found to be the lowest order spherically symmetric strain bubble solution of the equilibrium equations. This strain bubble consists of a standing strain wave 'core' of about 1.61×10^{-15} m (1.61 fm) radius, surrounded by a radially propagating phase wave field extending to infinity.

(i) The unique characteristic features of these radial strain wave fields manifest in the unique charge property of these strain bubbles. Detailed computations of strain energy content show that almost 65% of the total mass energy of the electron is contained in its core and the remaining 35% in its field.

(j) The nucleon core is represented by an important, lowest order, cylindrically symmetric solution of equilibrium equations. The proton consists of a positron entrapped inside a nucleon core through strong interaction and moving around the core center within a sphere of about 1 fm radius.

(k) The neutron consists of an electron entrapped inside the outer periphery of a proton through strong interaction and moving around the core center within a radius of about 2.6 fm.

(l) The strong interaction between two or more strain bubbles or particles is physically effected through the superposition of the strain fields in their cores. The range of strong interaction is small because the physical spread of the core region is small.

(m) Since the electrostatic wave fields of charge particles physically extend up to infinity, the range of electromagnetic interaction is said to be infinite. Similarly the gravitational interaction between two or more neutral particles is physically effected through the superposition of their neutral ripple wave fields, which too extends unto infinity.

In the grand maze of the unknown, the current paradigm has lead Fundamental Physics to a dead end.[11] To come out of this state, we need to change the direction of research and hence, change the current paradigm of Fundamental Physics. This book represents a bold attempt to show inadequacies of the current paradigm and to present a new paradigm of Fundamental Physics. The proposed paradigm shift is expected to enable significant advancements in our fundamental understanding of Nature in the near future.

REFERENCES

1. I. S. Sokolnikoff, "Tensor Analysis", John Wiley & Sons, Inc. (New York 1967)
2. Michelson, A. A.; Morley, E. W. (1887). "On the Relative Motion of the Earth and the Luminiferous Ether". http://www.aip.org/history/exhibits/gap/PDF/michelson.pdf
3. Comparision of "Old" and "New" Concepts: Reference Systems http://www.iers.org/documents/publications/tn/tn29/tn29_031.pdf
4. Albert Einstein, "On the Electrodynamics of Moving Bodies" http://www.fourmilab.ch/etexts/einstein/specrel/www/
5. Albert Einstein, "Relativity: The Special and General Theory" (December, 1916), http://www.bartleby.com/173/
6. I. S. Sokolnikoff, "Mathematical Theory of Elasticity", McGRAW-HILL Book Company, Inc. (New York 1956)
7. STANFORD ENCYCLOPEDIA OF PHILOSOPHY : Bohmian Mechanics http://plato.stanford.edu/entries/qm-bohm/
8. L. I. Schiff, "Quantum Mechanics", McGRAW-HILL Book Co., Inc. (Tokyo 1955)
9. Arthur Beiser, "Perspectives of Modern Physics", McGRAW-HILL BOOK Co. (Singapore 1981)
10. H. Margenau and G. M. Murphy, "The Mathematics of Physics and Chemistry", D. Van Nostrand Co. Inc. (New york 1962)
11. David Lindley, "The End of Physics", Basic Books, (New York 1993)

About the Author

With a humble background in the farming community of Punjab, India, the author completed his graduation in Mechanical Engineering from Punjab University in 1964. Immediately thereafter he joined the Indian Air Force, in the Aeronautical Engineering (Mechanical) branch and underwent a rigorous one year training course at Air Force Technical College, Bangalore in 1965. After a short stint at an operational air base, he was detailed for a short missile training course and then posted to an operational surface to air missile squadron in 1967. During his tenure in this missile squadron, he devoted his spare time for advanced studies and developed a special interest in the study of physics. It started with a firm belief that Physics, as the Mother of all sciences, is capable of logically explaining all natural phenomenon. For an in-depth understanding of physics, he had to study higher mathematics in parallel. Thus the study of physics and mathematics, for a fundamental level understanding of the natural phenomena and hence of Nature, became his life long quest.

In 1971, he was selected for a long technical staff course at Institute of Armament Technology, Pune (India). He topped in this course and won the running trophy for excellence. After a short stint at another operational air base, he was again selected in 1973 for a post graduation course in UK. He obtained his M. Sc. degree in the faculty of technology from the (then) Cranfield Institute of Technology (UK) with excellent academic performance. It was during this course that he first used computer aided mathematical modeling to solve practical problems. Later on he used this technique extensively during his 19 years of service tenure with Defense Research and Development Organization. During this time he managed to sustain his interest in the study, investigation and research in fundamental physics. But even after years of relentless study and reflection, he was unable to get any grasp of the workings of Nature at fundamental level. It was quite distressing for him to find that all this study could not help him visualize the shape, size or structure of elementary particles like the electron, photon or proton, neutron etc. Thereafter, he started probing the inadequacies in the current paradigm of fundamental Physics.

However, he finally realized the futility of pursuing the serious issue of fundamental research in physics by merely treating it as a hobby. He took premature retirement from service in 1993 and decided to devote his full time, energy and effort on this pursuit of fundamental research in Physics. The final outcome of this relentless effort was a new fundamental theory of space-time distortions as origin of elementary particles, which has been presented in this book.